SpringerBriefs in Environmental Science

SpringerBriefs in Environmental Science present concise summaries of cutting-edge research and practical applications across a wide spectrum of environmental fields, with fast turnaround time to publication. Featuring compact volumes of 50 to 125 pages, the series covers a range of content from professional to academic. Monographs of new material are considered for the SpringerBriefs in Environmental Science series.

Typical topics might include: a timely report of state-of-the-art analytical techniques, a bridge between new research results, as published in journal articles and a contextual literature review, a snapshot of a hot or emerging topic, an in-depth case study or technical example, a presentation of core concepts that students must understand in order to make independent contributions, best practices or protocols to be followed, a series of short case studies/debates highlighting a specific angle.

SpringerBriefs in Environmental Science allow authors to present their ideas and readers to absorb them with minimal time investment. Both solicited and unsolicited manuscripts are considered for publication.

More information about this series at http://www.springer.com/series/8868

Rama Kant Dubey • Vishal Tripathi
Ratna Prabha • Rajan Chaurasia
Dhananjaya Pratap Singh
Ch. Srinivasa Rao • Ali El-Keblawy
Purushothaman Chirakkuzhyil Abhilash

Unravelling the Soil Microbiome

Perspectives For Environmental
Sustainability

 Springer

Rama Kant Dubey
Institute of Environment & Sustainable
Development
Banaras Hindu University
Varanasi, Uttar Pradesh, India

Ratna Prabha
Chhattisgarh Swami Vivekananda
Technical University
Bhilai, Chhattisgarh, India

Dhananjaya Pratap Singh
ICAR-National Bureau of Agriculturally
Important Microorganisms
Mau Nath Bhanjan, Uttar Pradesh, India

Ali El-Keblawy
Department of Applied Biology
University of Sharjah
Sharjah, United Arab Emirates

Vishal Tripathi
Institute of Environment & Sustainable
Development
Banaras Hindu University
Varanasi, Uttar Pradesh, India

Rajan Chaurasia
Institute of Environment & Sustainable
Development
Banaras Hindu University
Varanasi, Uttar Pradesh, India

Ch. Srinivasa Rao
National Academy of Agricultural Research
Management
Hyderabad, Telangana, India

Purushothaman Chirakkuzhyil Abhilash
Institute of Environment & Sustainable
Development
Banaras Hindu University
Varanasi, Uttar Pradesh, India

ISSN 2191-5547   ISSN 2191-5555   (electronic)
SpringerBriefs in Environmental Science
ISBN 978-3-030-15515-5   ISBN 978-3-030-15516-2   (eBook)
https://doi.org/10.1007/978-3-030-15516-2

Library of Congress Control Number: 2019936174

This Springer imprint is published by the registered company Springer Nature Switzerland AG
The registered company address is: Gewerbestrasse 11, 6330 Cham, Switzerland

Preface

Soil is considered as a complex and dynamic biological system that provides diverse habitats for the microbial community and is a hotspot for belowground microbial interactions. The soil microbial community acts as a central factor in mediating the key ecosystem services and functions. However, unravelling this complex nature of the microbial world and successfully utilizing all positive interactions for multipurpose environmental benefits is still a major challenge for humanity. Although scientists have studied much about the microbially mediated soil functions such as biogeochemical cycling, enzyme activities, and soil respiration, still it is difficult to decipher the role of individual microbial species in key ecosystem functions as this requires high-throughput genetic, phylogenetic, and functional profiling of microorganisms from the complex soil matrix. However, the advent of next-generation molecular techniques has provided many inventive techniques for exploring microbial communities including the culture-independent study of the soil microbiome. Such cultivation-independent genomic approaches are more preferred for exploring the ecology and diversity of soil microbes because most soil microbes cannot be cultured. Moreover, soil metagenomics has deepened our knowledge about the complete characterization of new microorganisms, genes, gene products, and whole-community DNA sequences for community analysis and comparison of different microbial assemblages. Soil microbiome studies are not only necessary for elucidating the biological functions of soil but also for exploring the effect of various biotic and edaphic factors on soil microorganisms. Such studies provide better insight on how microbial communities are structured in soil, and therefore even the newer domains such as microbial biogeography are becoming increasingly popular in soil ecology.

With this background, the present book unveils the issues and challenges related to the exploration of soil microbial diversity, various kinds of interactions in the microbial world, and exploiting such beneficial interactions for ecosystem sustainability. Overall, the present book is aimed to provide state-of-the-art knowledge of modern techniques for characterizing soil microorganisms at taxonomic, functional, and community levels and also the frontiers in bioinformatics for unravelling the soil microbiome. We sincerely hope that this book will strengthen scientific understanding about various microbial interactions for improving soil fertility, restoring degraded systems, and also for augmenting the microbially mediated eco-system functions to meet the sustainable development goals. It shall be gratifying if this book can serve as a primer for graduate students and researchers in microbial ecology.

Varanasi, Uttar Pradesh, India   Rama Kant Dubey
Varanasi, Uttar Pradesh, India   Vishal Tripathi
Bhilai, Chhattisgarh, India   Ratna Prabha
Varanasi, Uttar Pradesh, India   Rajan Chaurasia
Mau Nath Bhanjan, Uttar Pradesh, India   Dhananjaya Pratap Singh
Hyderabad, Telangana, India   Ch. Srinivasa Rao
Sharjah, United Arab Emirates   Ali El-Keblawy
Varanasi, Uttar Pradesh, India   Purushothaman Chirkkuzhyil Abhilash

Acknowledgments

We sincerely wish to thank Margaret Deignan, Jill Ritchie, Susan Westendorf and Chandhini Kuppusamy from Springer for their editorial support, guidance, and cooperation. Rajan Chaurasiya is grateful to CSIR for a Junior Research Fellowship. P.C. Abhilash is grateful to ICAR for the Lal Bhadur Shastri Outstanding Young Scientist Award in Natural Resource Management. Dhananjaya Pratap Singh and Cherukumalli Srinivasa Rao are grateful to ICAR for financial support. Special thanks go to Prof. Panjab Singh, The President, National Academy of Agricultural Sciences (NAAS) for his continuous motivation and encouragement.

Abstract

The soil is one of the critical resources of planet Earth, maintaining biodiversity and ecosystem services. It contains the largest microbial diversity, which is crucial in governing the functions of the soil. Therefore, proper elucidation of soil microbial diversity is important to understand the critical factors contributing to soil ecological services and health. Still, very little has been known about soil microbial diversity as only a small fraction of soil microorganisms have been cultured and studied for their structural and functional attributes. Because only about 1% of microbes are culturable, the uncultured microbiota is considered as a treasure trove of the microbial world. Therefore, exploring this hidden resource is important, especially for agricultural and environmental microbiology. The recent progress in molecular microbial ecology has provided powerful tools to explore the genetic information of soil microorganisms. Omic technologies (such as quantitative PCR, differential gradient gel electrophoresis, metagenomics, metatranscriptomics, metaproteomics, and single-cell genomics) are the most powerful tools for exploring the complex aboveground–belowground microbial diversity of the soil and understanding the structural and functional attributes of this diversity. This book depicts various conventional and advanced technologies for exploring soil microbial diversity to understand its utility and importance in soil functions and ecosystem services, and, most importantly, achieving sustainable development goals.

Keywords Biodiversity · Ecosystem services · Land restoration · Metagenomics · Metatranscriptomics · Metaproteomics · Microbial ecology · Soil microbiome · Single-cell genomics · Sustainable agriculture

Abbreviations

ACC	1-Aminocyclopropane-1-carboxylate
ACCD	1-Aminocyclopropane-1-carboxylate deaminase
AMF	Arbuscular mycorrhizal fungi
ARISA	Automated ribosomal intergenic spacer analysis
BLAST	Basic local alignment search tool
CARD	Catalysed reporter deposition
CDSs	Coding sequences
CH_4	Methane
CLP	Cyclic lipopeptides
CNVs	Copy number variants
CO_2	Carbon dioxide
COGs	Clusters of orthologs groups
DDT	Dichlorodiphenyltrichloroethane
DGGE	Denaturing gradient gel electrophoresis
DNA	Deoxyribonucleic acid
DTT	Dithiothreitol
FACS	Fluorescence-activated cell sorting
FAME	Fatty acid methyl esters
FASP	Filter-aided sample preparation
FGA	Functional gene arrays
FISH	Fluorescence in situ hybridization
GC-MS	Gas chromatography–mass spectrometry
GHGs	Greenhouse gases
GOLD	Genomes Online Database
GOS	Global Ocean Sample
HCH	Hexachlorocyclohexane
IAA	Indole acetic acid
ITS	Internal transcribed spacer
KEGG	Kyoto Encyclopedia of Genes and Genomes
LC	Liquid chromatography
LC-MS	Liquid chromatography–mass spectrometry

MDA	Multiple displacement amplification
N$_2$O	Nitrous oxide
NGS	Next-generation sequencing
OTC	Open top chamber
OTU	Operational taxonomic unit
PAHs	Polycyclic aromatic hydrocarbons
PCB	Polychlorinated biphenyl
PCR	Polymerase chain reaction
PGPF	Plant growth-promoting fungi
PGPR	Plant growth-promoting rhizobacteria
PLFA	Phospholipid fatty acid analysis
Q-PCR	Quantitative polymerase chain reaction
Q-TOF-MS	Quadrupole-time of flight-MS
RDP	Ribosomal Database Project
RNA	Ribonucleic acid
RT-PCR	Reverse transcriptase polymerase chain reaction
SCG	Single-cell genomics
SDS	Sodium dodecyl sulfate
SIP	Stable isotope probing
SNPs	Single-nucleotide polymorphisms
SNVs	Single-nucleotide variants
SOM	Soil organic matter
SSCP	Single-strand conformation polymorphism
SSU	16S small subunit
TGGE	Temperature gradient gel electrophoresis
TPH	Total petroleum hydrocarbon
T-RFLP	Terminal restriction fragment length polymorphism
TRFs	Terminal restriction fragments
VOCs	Volatile organic compounds
WGA	Whole genome amplification

Contents

1   **Introduction**   1
     1.1   Unexplored Soil Microbial World: A Solution for the Multiple
          Challenges   1
     1.2   Microorganisms Under a Changing Climate   3

2   **Belowground Microbial Communities: Key Players for Soil and
     Environmental Sustainability**   5
     2.1   Microorganisms for Agriculture and Environmental
          Remediation   5
     2.2   Plant Growth-Promoting Microorganisms (PGPMs).   6
     2.3   Plant Growth-Promoting Traits and Microbial Functions   8
     2.4   Microbes Improve Agricultural Production and Nutritional
          Quality   10
     2.5   Biocontrol Agents and Resistance Against Plant Diseases   11
     2.6   Microorganisms Improve Soil Quality.   12
     2.7   Microbially Assisted Extensification for Improving
          Agricultural Production from Degraded Lands   15
     2.8   Microorganisms for Phyto-Bioremediation, Carbon
          Sequestration, Biomass, and Bioenergy Production   16
     2.9   Major Challenges for Wide-Scale Utilization of
          Microbial Services   17

3   **Methods for Exploring Soil Microbial Diversity**   23
     3.1   Phospholipid Fatty Acid Analysis (PLFA).   24
     3.2   Fluorescence In Situ Hybridization (FISH)   25
     3.3   Denaturing Gradient Gel Electrophoresis (DGGE)   26
     3.4   Terminal Restriction Fragment Length Polymorphism
          (T-RFLP)   27
     3.5   Automated Version of RISA (Ribosomal Intergenic
          Spacer Analysis (ARISA)   28
     3.6   Single-Strand Conformation Polymorphism (SSCP).   28
     3.7   Stable Isotope Probing (SIP)   30

| | 3.8 | Quantitative PCR (Q-PCR) | 30 |
| | 3.9 | DNA Microarray | 31 |

4   Single-Cell Genomics and Metagenomics for Microbial Diversity Analysis 33

	4.1	Single-Cell Genomics (SCG)	34
		4.1.1 Cell Isolation	36
		4.1.2 Whole-Genome Amplification	37
		4.1.3 Interrogation of WGA Products	37
		4.1.4 Overview of Single-Cell Sequencing Errors	38
	4.2	Metagenomics	38
		4.2.1 Metagenomics and Soil Microbial Diversity	39
		4.2.2 Metagenomics Data Analysis: Approaches and Challenges	41
		4.2.3 Metagenomics Data Analysis: Available Tools and Techniques	44
		4.2.4 Metagenomics: Success Stories	48

5   Metatranscriptomics and Metaproteomics for Microbial Communities Profiling 51

	5.1	Metatranscriptomics	51
		5.1.1 Isolation and Processing of Microbiome mRNA	53
		5.1.2 Computational Analysis of Metatranscriptomics Data	53
		5.1.3 Metatranscriptomics and Soil Microbial Diversity	54
	5.2	Metaproteomics	55
		5.2.1 Metaproteomics: Benefits and Challenges	56
		5.2.2 Metaproteomics Approaches	57
		5.2.3 Metaproteomics and Soil Microbial Communities	59

6   Bioinformatics Tools for Soil Microbiome Analysis 61

	6.1	Tools for Assembly and Annotation	61
	6.2	Tools for Taxonomic Profiling	66
	6.3	Tools for Functional Profiling	68
	6.4	Tools for Underlying Interactome	68
	6.5	Tools for Statistical Tests	68
	6.6	Simulators Tools	69
	6.7	Tools for Single-Cell Sequencing Analysis	69
	6.8	General Toolkits	70

7   Conclusion and Future Perspectives 71

| | 7.1 | Conclusion | 71 |
| | 7.2 | Future Microbiome Research Directions: How Do They Engage Themselves? | 73 |

References 77

Index 101

About the Authors

Rama Kant Dubey is a doctoral research fellow, working in the area of plant–microbe interactions, sustainable agriculture, and soil system sustainability at the Institute of Environment & Sustainable Development, Banaras Hindu University, India. Being a Green Talent Awardee (BMBF, Govt. of Germany), he has also worked as a Guest Researcher at Helmholtz Zentrum München, Neuherberg, Germany. He received his M.Sc. in Industrial Microbiology from Devi Ahilya Vishwavidyalaya, Indore. He has received the AU-CBT Excellence Award of Biotech Research Society of India and NASI-Swarna Jayanti Puraskar from the National Academy of Sciences, India. Mr. Dubey has been serving as an active member for the Commission on Ecosystem Management (CEM), IUCN, and serving as an Editor of the 'Agroecosystems Newsletter' of the IUCN-CEM Agroecosystems Specialist Group and special issue in Agronomy (MDPI). He is also serving as an expert reviewer for various international research journals from Elsevier, Springer-Nature, Wiley, and PLOS.

Vishal Tripathi is a Research Scholar at the Institute of Environment and Sustainable Development, Banaras Hindu University. His research interests lie in restoration and management of polluted lands and understanding the plant–microbe–pollutant interactions under the changing climate to develop future remediation strategies. He has been a recipient of the prestigious Green Talents Award of the Federal Ministry of Education & Research (BMBF), Germany for the year 2018. He has also been honored with the AU-CBT Research Excellence award of The Biotech Research Society of India, for the year 2016. He is a member of the Commission on Ecosystem Management (CEM) and serves as an expert reviewer for several international journals, namely, *Scientific Reports* (Nature Publishing Group), *Biomass and Bioenergy* (Elsevier), *Agronomy for Sustainable Development* (Springer), *Restoration Ecology* (Wiley), *Biodegradation* (Springer), *Environmental Management* (Springer), *PLOS One* (PLOS), and *Energy, Ecology & Environment* (Springer).

Ratna Prabha is a SERB-National Postdoctoral Fellow (DST, Govt. of India) at Chhattisgarh Swami Vivekanda Technical University, Bhilai. She obtained her master's degree in Bioinformatics from Banasthali Vidyapeeth and her Ph.D. in Biotechnology from Mewar University, India. She has been engaged in developing various digital databases on plants and microbes and has published two edited books, many book chapters, and various research papers and review articles in journals of international repute. Her current research interest lies in microbe-mediated stress management in plants, database development, comparative microbial genome analysis, phylogenomics and pangenome analysis of prokaryotic genomes, and metagenomics data analysis. She has completed several bioinformatics demonstration tasks at different national training programs on bioinformatics and computational biology. She has been awarded the Young Scientist Award at G.B. Pant University of Agriculture and Technology; S&T SIRI, Telangana; and CGCOST, Chhattisgarh.

Rajan Chaurasia is a Research Fellow (CSIR-NET-JRF) at the Institute of Environmental and Sustainable Development, Banaras Hindu University. His research interests are crop diversification, emerging cropping patterns, plant–microbe interactions, microbial diversity, and sustainable agriculture. He has received the Young Scientist Travel Award from the International Union of Biological Sciences (IUBS), Paris. He is a life member of the Association of Environmental Analytical Chemistry of India (AEACI), the Society for Science of Climate Change and Environmental Sustainability and an active member of the Global Land Programme (GLP), Global Soil Biodiversity Initiative (GSBI), and the Commission on Ecosystem Management of IUCN.

Dhananjaya Pratap Singh is a Principal Scientist in biotechnology at ICAR-National Bureau of Agriculturally Important Microorganisms (NBAIM), Maunath Bhanjan, India. He obtained his master's degree from G.B. Pant University of Agriculture and Technology, Pantnagar, and his Ph.D. in Biotechnology from Banaras Hindu University, Varanasi. His research interests include plant–microbe interactions, bio-prospecting of metabolites of microbial and plant origin, microbe-mediated stress management in plants, metabolomics-driven search for small molecules, and bioinformatics in microbial research. He was involved in the development of a supercomputational facility for agricultural bioinformatics in the microbial domain at ICAR-NBAIM under the National Agricultural Bioinformatics Grid (NABG) program of ICAR. He has been awarded with various prestigious awards including the Dr. A.P.J. Abdul Kalam Award for Scientific Excellence in 2016 by Marina Labs. He has more than 134 publications and one Indian patent to his credit.

Ch. Srinivasa Rao is the Director of the ICAR-National Academy of Agricultural Research & Management (NAARM), Hyderabad. He received his M.Sc. (Ag) from ANGRAU, Bapatla, AP, his Ph.D. from IARI, New Delhi, and did Post-Doctoral research at Tel-Aviv University, Israel. His specializations are soil carbon sequestration, climate change, agriculture contingency planning, conservation policy, and agriculture research management. He was coordinator for AICRP on dryland agriculture and Director of CRIDA, Hyderabad during 2013–2017. He was the National Coordinator for the ICAR-flagship program on climate change, NICRA, and served as the technical chairman of the National Mission for Sustainable Agriculture, a climate change negotiator in the Indian delegation represented at UNFCCC, SBSTRA, and COP meetings, etc. He is executive board member of the International Dryland Development Commission and Member, Asian Carbon Network. He is a Fellow of NAAS, ISSS, ISPRD, and received prestigious awards from the Honourable President and Prime Minister of India. He has published 255 research papers, 25 books, and serves as a regular reviewer for 15 international journals.

Ali El-Keblawy is a Professor and Chair of the Applied Biology Department, University of Sharjah, UAE. He was awarded his Ph.D. degree in Plant Ecology from Tanat University, Egypt and Windsor (Ontario, Canada) in a joint program. Dr. El-Keblawy has established the Sharjah Seed Bank and Herbarium and was its director for 5 years. His research interests include plant ecology, biodiversity and conservation of desert plants, invasion biology, rangeland management, propagation of native plants of the Arab Gulf deserts, and domestication of desert native plants for urban landscaping. He has been awarded the Distinguished Faculty award in Scientific Research, University of Sharjah.

Purushothaman Chirakkuzhyil Abhilash is a senior Assistant Professor of sustainability science in the Institute of Environment & Sustainable Development at Banaras Hindu University in Varanasi, India and Chair of the Agroecosystem Specialist Group of IUCN-Commission on Ecosystem Management. He is a Fellow of the National Academy of Agricultural Sciences (NAAS), and his research interests include land degradation and restoration, bioremediation, integrated remediation techniques, and harnessing plant–microbe interactions as low-input techniques for multipurpose environmental benefits. He is serving on the editorial board of *Agronomy*, *Biodegradation*, *Biomass & Bioenergy*, *Energy, Ecology & Environment*, *Environmental Management*, *Land Degradation & Development*, *Land*, *Restoration Ecology, Sustainable Earth,* and *Tropical Ecology* and as an expert of IPBES, IPCC, UNDP-BES Network, UNCCD, and IRP of UNEP.

Chapter 1
Introduction

Abstract Soil microorganisms play a vital role in soil functions influencing the biogeochemical cycle, soil fertility, plant health, and aboveground ecosystems. The soil harbours more diverse communities of microorganisms than any other environmental component. We have a narrow understanding of how microbial diversity regulates soil functioning and in turn affects ecosystem sustainability. Studies of soil microorganism-mediated processes responsible for soil functions have largely been neglected. With increasing pressure on soils to meet the demands of the rapidly increasing human population for food, fodder, fibre, biofuel, timber, clean water, etc., it is imperative that research in soil microbiology focuses on the structure and functions of the soil microorganisms to delineate microbe-mediated soil processes and optimise them for enhanced production and better soil function. Moreover, soil also acts as a reservoir of carbon because its soil carbon sequestration potential helps in reducing atmospheric CO_2 levels. However, it is believed that warming climate conditions in the changing climate can negatively affect the carbon sequestration potential and other functions of the soil. To negate the climate impacts arising from increased CO_2 emissions from the soil, it is essential that we have a deep understanding of the processes of soil carbon storage. As the microbial activities in the soil largely regulate its functions, including soil carbon sequestration, it is important to gain deeper insights into the soil microbial world to address the issues of climate change and food security.

Keywords Soil microorganisms · Soil functions · Ecosystem sustainability · Carbon sequestration · Climate change · Food security

1.1 Unexplored Soil Microbial World: A Solution for the Multiple Challenges

Our rapidly and ever-increasing burgeoning population is projected to be 8.5 billion in 2025 (FAO 2013). At least a 50% increase in food production is needed to balance the nutrition of 1 billion undernourished and 1 billion malnourished people globally. Sustainable approaches for improving agricultural production and restoring degraded soil are necessary to meet global food and nutritional security concerns with better environmental sustainability, which is also the target of the sustainable development goals (SDGs) (Abhilash et al. 2016a; Singh et al. 2018). With increasing crop yield, we have to reduce adverse effects on climate, human health, aquatic ecosystems, biodiversity, soil systems, and all ecosystem services (Power 2010). Along with the burden of increasing food productivity, growing urbanization, industrialization, warming climate, and agrochemical pollution also place extraneous pressure on the agricultural production system. Thus, there is a need to develop and adopt sustainable methods for food production with improved soil quality, ecosystem resilience, increased crop yield, and nutritional content with minimum environmental risks (Singh et al. 2018). To increase food production with the existing agricultural land, we need to explore options for the promotion of better agronomic practices, adoption of genetically improved varieties of crops, and the function of belowground microbial communities in strengthening plant–microbial interactions and ecosystem functioning (Dubey et al. 2016b).

Soils are the hub for maintaining all kinds of ecosystem services and also provide a key resource for food, feed, fiber, and energy production. Soil has the highest level of microbial diversity compared to any other environment. The soil can harbour 1 million distinct genomes per gram, belonging to 4,000 to 10,000 different microbial species, which constitute 60% of the total Earth biomass (Torsvik and Overeas 2002; Singh et al. 2009), but the study of soil microbial communities and functions is still in its infancy. Microbial populations inhabiting the soil have enormous genomic, proteomic, and metabolomic diversity that reflect huge functional attributes of direct environmental concerns. It is reported that approximately 10^{30} bacteria are present on our Earth. On an average, 1 g soil harbours about 10^9 bacteria, 10^8 actinomycetes, 10^6 fungal cells, and, in the cumulative figure, 10,000 to 50,000 microbial species (Roesch et al. 2007). These microbes support various soil functions, such as nutrient cycling, soil health and fertility, plant health and productivity, bioremediation and antibiotic resistance, and make the soils a living body (Torsvik and Overeas 2002; Singh and Trivedi 2017). Microbial diversity of the soils varies across different geographic locations because of such abiotic and biotic factors as temperature, precipitation, vegetation, soil structure, and space and time interval. Structural shifts in the microbial community are always linked with the changes in its functional attributes that drive the agro-ecosystem (Dubey et al. 2015; Tripathi et al. 2015a).

Chemical fertilizers are used profusely in agro-ecosystems to enhance agricultural production through providing essential nutrients such as nitrogen (N), phosphorus (P), and potassium (K) to crops. However, excessive use of synthetic fertilizers has shown harmful effects on the environment and crop productivity, contamination of the belowground water table, soil nutrient surface runoff, aquatic ecosystem eutrophication, crop susceptibility to diseases, and ultimately loss in agro-economy (Abhilash et al. 2013a). Furthermore, heavy dependence on synthetic agrochemicals results in the loss of soil fertility and variable impacts on the composition and functions of soil microbiota. In addition, chemical fertilizers have often

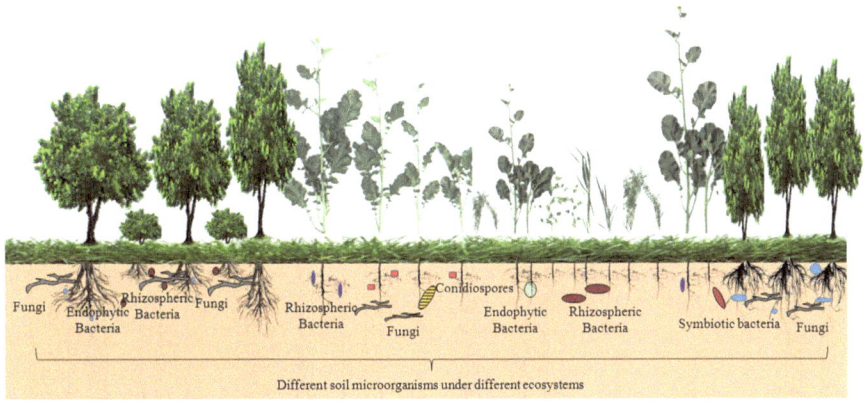

Fig. 1.1 A peek inside the belowground microbial world under the soil. Soils harbour a huge unexplored diversity of bacteria, fungi, and actinomycetes, etc.

low nutrient use-efficiency, and only a small fraction of the applied nutrients are assimilated by the plants (Adesemoye and Kloepper 2009). Therefore, it is imperative to seek effective and enduring solutions for providing food security through sustainable and environmentally friendly biological options based on the establishment of natural agro-ecological ways. Thus, the purposeful utilization of beneficial soil microbes for plant growth promotion and biocontrol in agriculture, remediation and restoration of degraded lands, production of biomass, and bioenergy feedstock could be attractive options to address the worldwide problems arising from use of synthetic pesticides and chemical fertilizers (Akinsemolu 2018) (Fig. 1.1).

1.2 Microorganisms Under a Changing Climate

There exists an invisible, unseen, and complex microbial life and interactome beneath our feet. Microbes exist everywhere, including in all extreme environments, where they are called extremophiles (Rothschild and Mancinelli 2001). Recent study suggests that the seeds are microbiologically active and these beneficial seed microbiomes have the capacity to improve crop stress tolerance and productivity (Berg and Raaijmakers 2018). Therefore, exploration of a single microbial species with contemporary activity would be as fascinating as the invention of a star (Jansson 2013). Unique microbial diversity such as chasmoliths, hypoliths, and endoliths having threatening challenges to life were reported from the antarctic soils of McMurdo Dry Valleys, a hyper-polar arid desert (Pointing et al. 2010). Soil microorganisms in the extreme dryland are responsible for exchanging gaseous emissions, maintaining the biogeochemical cycle, and regulating bioweathering by epilithic and endolithic communities in sub-arid or high-arid conditions, respectively. Stability of the soil in such ecosystems is governed via hypolithic biomass and associated extracellular polymeric substances (Pointing and Belnap 2012). Different N-fixing, denitrifying, decomposing, and P-acquisitioning microbes, which regulate nutrient cycling, occur in the terrestrial ecosystem. These abundant and diverse microfloras also aid the process of mineralization in the soils and productivity of the crop plants (van der Heijden et al. 2008). Genomic and proteomic analysis of psychrophiles reveals the secret of its survival in extreme permafrost soil. The mechanism behind the adaptation is overexpression of cold-shock proteins that critically regulate protein folding and three-dimensional structures. Gram-negative α-, β-, and γ-proteobacteria (*Pseudomonas* sp., *Vibrio* sp.), *Flavobacterium*, and gram-positive bacteria (*Micrococcus* sp., *Arthrobactor* sp.) are the dominant microorganisms present under permafrost soil (Amico et al. 2006). Morris reported that 68.5% of bacteria are microaerobic in nature, having a high affinity for the cytochrome oxidase gene. Shotgun metagenomic results showed a high percentage of these bacteria in Puerto Rican rainforest soils in comparison to Michigan agricultural soils and Michigan deciduous forest soils. Still, the unravelling of the processes and factors responsible for the abundance of micro-oxic bacteria in Puerto Rican rainforest soil is a topic of deep and focused research. Apart from the huge soil microbial diversity, its utilization in addressing the global problems of food, feed, nutrition, and energy security is challenging and need special attention. For the problems arising from dependency on chemical fertilizers and fertilizers, N-N-N-rehabilitation strategies based on maintaining microbial flora in the soils could be viable options (Canfield et al. 2010).

Microbial communities in the soils govern fertility, physicochemical properties, and balance in ecosystem functioning, which is an essential issue in the changing climatic era. Global average temperature and greenhouse gas emissions are increasing steadily and, in such conditions, aridity of the soil may increase its contribution to global CO_2, CH_4, and N_2O emissions from the terrestrial agro-ecosystem (Dubey et al. 2016b). To cope up with such issues, above- and belowground agriculturally important microbial communities need to be explored and utilized as necessary (Abhilash and Dubey 2014; Abhilash et al. 2015; Tripathi et al. 2016b). Moreover, the response of microorganisms to increasing temperature and elevated concentrations of CO_2 could be essential studies that may reflect shifts in the communities of Proteobacteria, Firmicutes, Actinobacteria, Bacteroidetes, and Acidobacteria caused by elevated CO_2 concentration (He et al. 2012). For better soil management and achieving the targets of SDGs, proper understanding about the structure and functions of microbial diversity and its inter- and intra-communications is essential to regulate soil processes and ecological functioning.

Chapter 2
Belowground Microbial Communities: Key Players for Soil and Environmental Sustainability

Abstract One of the key functions of soil microorganisms is to promote plant health and increase soil productivity. Some indigenous microorganisms of contaminated soil systems also have the capability to degrade the soil contaminants, and thus are frequently used in bioremediation purposes. Most of these microbes belong to the category of plant growth-promoting microorganisms, which perform various functions including providing nutrients to the plants, conferring disease resistance, and combating temperature, salinity, and other abiotic stresses. These microbes also support the growth of plants in degraded and contaminated soil systems, aiding phyto-remediation. In coming decades, food security and climate change are expected to be the most serious problems for the planet Earth. Plant growth-promoting microorganisms can provide a sustainable solution to these problems by increasing the production of crop plants, and reducing the use of chemical fertilizers and pesticides, thus decreasing agricultural pollution. The enormous diversity of soil microorganisms gives an excellent opportunity for exploring new plant growth-promoting rhizobacteria; however, only 1% of the total soil microorganisms are culturable in nature, so the majority of the promising strains remain unexplored. The advent of the novel 'omic' technologies provides an excellent opportunity to harness this potential of the belowground microbial world.

Keywords Bioremediation · Climate change · Food security · Stress tolerance · Sustainable agriculture · Plant growth-promoting microorganisms

2.1 Microorganisms for Agriculture and Environmental Remediation

Microorganisms have a very important role in sustainable agriculture as well as in the field of environmental cleanup technologies. Microbes also have the potential to enhance global sustainability through achieving the sustainable development goals (Akinsemolu 2018). Further details about their role and activity in the mentioned areas are discussed next (Fig. 2.1).

© The Author(s), under exclusive license to Springer Nature Switzerland AG 2020   5
R. K. Dubey et al., *Unravelling the Soil Microbiome*, SpringerBriefs in
Environmental Science, https://doi.org/10.1007/978-3-030-15516-2_2

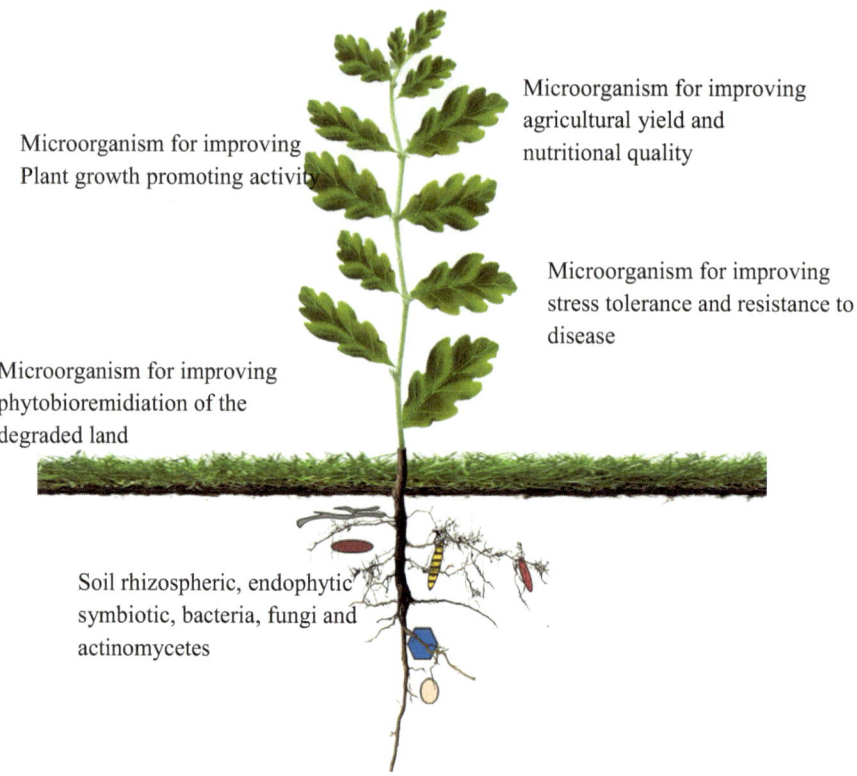

Microorganism for improving agricultural yield and nutritional quality

Microorganism for improving Plant growth promoting activity

Microorganism for improving stress tolerance and resistance to disease

Microorganism for improving phytobioremidiation of the degraded land

Soil rhizospheric, endophytic symbiotic, bacteria, fungi and actinomycetes

Fig. 2.1 The importance of soil microorganisms and their functions. Soil microbial communities play a key role in a myriad of processes such as improving soil quality and plant health and productivity along with degraded land restoration and fostering ecosystem services

2.2   Plant Growth-Promoting Microorganisms (PGPMs)

Plant growth-stimulating microbes, reported from various natural, biotic, and abiotic stressed soils, have the potential to enhance plant biomass and development and the yield of cereals, legumes, oil seeds, and vegetable crops (Bashan et al. 2012; Nadeem et al. 2014; Abhilash et al. 2016a, b; Akinsemolu 2018). On the basis of their habitat, plant growth-stimulating microbes can be categorised as rhizospheric, those that colonize the plant rhizosphere and rhizoplane (root surface), and endophytes, which colonize inside the plant roots, stems, leaves, or seeds without inducing any deleterious impact on associated plants. These microbes may enhance the plant performance via increased germination, growth, roots and shoot length, yield, content of chlorophyll, N, protein and antioxidants, hydraulic activity, and increase tolerance against various stresses (Nadeem et al. 2014). Although adapted to different environments, all soil microorganisms, including plant growth-promoting rhizobacteria (PGPR) and plant growth-promoting fungi (PGPF), utilize almost similar mechanisms for promoting plant growth (Glick 2012). However, the functioning and proliferation of rhizospheric bacteria may be influenced by the alteration in the soil pH, moisture, temperature, nutrient status, and the presence of competing soil bacteria (Santoyo et al. 2016).

PGPRs belong to the diverse group of bacterial strains that reside in the rhizospheric region, at the plant–soil interface, and can stimulate plant growth and development in a variety of ways (Dubey et al. 2017). Many gram-positive and gram-negative plant growth-stimulating bacteria have been found to colonize in the plant rhizosphere and confer beneficial effects that can be correlated with their ability to form biofilm, chemotaxis, and production of exopolysaccharides, phytohormones, and 1-aminocyclopropane-1-carboxylate (ACC) deaminase (Nautiyal et al. 2013). Plant growth-stimulating bacteria enhance plant biomass and health through distinct processes such as plant growth enhancement, improved nutrient use efficiency, and stress tolerance. In general, phytostimulators are known to improve plant growth by producing the myriads of phytohormones such as auxins, cytokinins, and gibberellins. Secretion of indole-3-acetic acid (IAA) and ACC deaminase are considered as direct mechanisms (Glick 2012). ACC deaminase inhibits ethylene synthesis by degrading the ACC (a precursor of ethylene) in plants, which becomes active against the various biotic (bacteria, fungus, insects, viral strains) and abiotic (high temperature, drought, flood, salinity, heavy metal, pesticide contamination) stresses (Glick 2014). In high concentrations, ethylene can lead to plant growth inhibition or even death. Biofertilizer bacterial strains fix N, solubilize P, mobilize K, and chelate with Fe and Zn to form siderophores. Biocontrol agents fight against phytopathogens and save the plants from diseases by nutrient competition, induced systemic resistance, and antimicrobial chemicals (Dubey et al. 2017). The following genera of PGPR, such as *Azotobacter, Azospirillum, Bacillus, Burkholderia, Chromobacterium, Erwinia, Enterobacter, Flavobacterium, Klebsiella, Micrococcus, Rhizobium, Pantoea,* and *Pseudomonas*, as well as *Serratia*, are well-known biocontrol agents with the potential to protect the plants from myriads of phytopathogens (Glick 2012; FigueroaLópez et al. 2016).

Endophytes are microorganisms that colonize host tissues and establish a relationship wherein both partners obtain benefits from their interactions. Some microbes inhabit within the plants in the form of endophytes but do not harm them by establishing symbiotic, mutualistic, commensalistic, and trophobiotic relationships (Bertani et al. 2016). The long-term co-evolution of plants and endophytic bacteria has resulted in an intimate ecosystem that helped plants to adapt and survive in multiple (biotic and abiotic) stress situations and enhanced the ecological balance of the natural ecosystem. Endophytic bacteria start inhabiting host plant tissues mainly by the root system; however, inflorescences, shoots, and sometimes cotyledons are also reported as a route for entry. Once inside the plant, bacteria can colonize at the primary tissue for entry as well as systemically throughout the plant (Bulgarelli et al. 2013). Endophytic bacteria have one main advantage over rhizospheric bacteria in that once these are established inside the plant tissues, they are no longer subject to fluctuating soils conditions. Endophytic bacteria colonize interior tissues by the production of cellulolytic enzymes and motility. Studies on bacterial endophytes may facilitate understanding about their interactions with plants and associated biotechnological applications for phytoremediation, rhizoremediation, soil conditioning, and remediation of soil contaminants. Plant growth-promoting endophytes perform N-fixation, mineral solubilization, phytohormone secretion, produce ammonia, hydrogen cyanide, siderophores, and ACC-deaminase enzyme, and provide synthesis of surface-active compounds such as biosurfactants and antagonists against plant pathogens. They have the great biochemical potential of producing hydrolytic enzymes that exhibit natural competence for xenobiotic degradation during phytoremediation (Li et al. 2012). These organisms are generally isolated from various plants for their identification and functional characterization, population dynamics, and diversity studies use as microbial inoculants to improve plant health and as a source of distinct secondary metabolites (Potshangbam et al. 2017).

For the selection of potential plant growth-promoting bacteria (PGPB), the isolates have to be first isolated, identified morpho-physiologically and taxonomically, and preliminarily characterized on the basis of in vitro biochemical analysis. For molecular taxonomic establishment, 16S rRNA-based analysis is the highly preferred method to identify genera and species such as *Bacillus*, *Rhizobium*, *Burkholderia*, *Enterobacter*, *Pantoea*, and *Serratia* (Szilagyi-Zacchin et al. 2014).

2.3 Plant Growth-Promoting Traits and Microbial Functions

The positive partnership of the plant–microbe system helps in plant growth and biomass stimulation. The process of biofertilization enhances nutrient uptake in the plants, as is evident from the studies on cereals, oilseeds, vegetables, and other edible crops. Increase in the dry weight of grains and whole plants in maize was observed when the plants were treated with the PGPR *Burkholderia cepacia*, *Azospirillum brasiliense*, and *Herbaspirillum seropedicae* individually for comparative analysis of plant growth in N-deprived soils (Pérez-Montaño et al. 2014). Field release of *Pseudomonas fluorescens* DR54 enhanced maize growth and the soil phosphorus pool (Krey et al. 2013). Application of *P. fluorescens* DR54 enhanced the colonization of AM fungi that subsequently increased the fine root hairs of maize by 30% and hence, enhanced phosphate mobilization under P-deprived soils was observed. The height of the inoculated maize plant was enhanced by 11 cm and 12 cm from the control in the first and second tested years, respectively. The performance of two- and three-component microbial inoculants, that is, *Pseudomonas fluorescence* F113 and consortia 3 *Glomus* isolates, and bivalent consortia with either *Azospirillum lipoferum* CRT1 or *A. brasilense* UAP-154 or CFN-535 was evaluated in maize (Couillerot et al. 2013). After 10 days of inoculation with two-component consortia, the enhancement in the shoot dry weight was about 0.04 g/plant whereas after 21 days it was about 0.17 g/plant.

Furthermore, post-bacterial (*Herbaspirillum seropedicae* SmR1) inoculation performance of different wheat genotypes has shown that the roots of different cultivars of wheat viz. CD 119 and 120 responded positively when inoculated with the strain SmR1 (Neiverth et al. 2014). Increase in the root fresh weight leaf fresh and dry weight in the CD 119 cultivar by 0.32 g, 1.06 g, and 0.72 g, respectively, was observed compared to the control within 7 days. Similarly, there has been increase in cultivar CD 120 by 0.11 g, 0.28 g, and 0.06 g, respectively, from the control. The productivity of the genotype CD 120 was significantly enhanced after the inoculation of SmR1 in the absence of urea (Neiverth et al. 2014). Ramesh et al. (2014) also reported improved wheat growth treated with the zinc-solubilizing *Bacillus aryabhattai*, which included three strains, namely, MDSR7, MDSR11, and MDSR14, which enhanced plant height by 17 cm, 12 cm, and 22 cm, respectively. Substantial improvement in the shoot dry weight (2.3 g/plant, 0.6 g/plant, and 2.7 g/plant) was also observed with MDSR7, MDSR11, and MDSR14 strains, respectively. In the case of rice, a key staple food, inoculation of the cyanobacterium *Calothrix elenkinii* improved growth of shoot length by about 12.7 cm and root length by 7.37 cm compared to the control. Fresh and dry weight of rice plants were also enhanced about 323.33 mg and 24.67 mg, respectively, in comparison to control, which was 76.67 and 15.10 mg, respectively. The physiological attributes such as chlorophyll (up to 6.21 µg/g in leaf and 2.66 µg/g in root), IAA production (26.95 µg/g in leaf and 14.07 µg/g in root), nitrogenase activity (11.11 nmol C_2H_4/plant/h), and hydrolytic and defence enzymes were also enhanced (Priya et al. 2015). Among hydrolytic enzymes, chitosanase activity was high in inoculated plants, by 10.7 and 9.0 IU/mg fresh weight in the root and shoot, respectively, compared to the control (Priya et al. 2015). Overall, cyanobacterial inoculation helped the root system against any fungal infections and subsequently promoted its growth. Similar observations of enhanced agronomic, physicochemical, enzymatic, and biochemical parameters were also observed in rice cultivars inoculated with three cyanobacterial strains (Singh et al. 2011). Inoculation of cyanobacterial strains also enhanced accumulation of phenylpropanoid metabolites, especially phenolic acids and flavonoids in the rice plants that were correlated with the enhanced antioxidant properties under stress condition. In another study, five endophytic diazotrophic bacterial strains (KW7-S22, KW7-S06, SW521-L21, CB-R05, HS-R01) inoculated in rice enhanced the length of the fresh leaf (39.11, 41.29, 37.99, 50.25, and 43.21 cm, respectively) and root (21.98, 25.56, 19.41, 33.73, and 28.11 cm, respectively). The length of dry leaf and root was also enhanced (0.36, 0.39, 0.31, 0.49, and 0.42 cm) and (0.24, 0.27, 0.18, 0.39, and 0.33 cm, respectively) after treatment of all five strains (Ji et al. 2014). Inoculation of *Bacillus aryabhattai* strains MDSR7, MDSR11, and MDSR14 in soybean enhanced the growth of the plants. Plant height was enhanced to 38, 32, and 40 cm respectively, in the three strains, as compared to the control which was about 24 cm. Similarly, the shoot and root dry weight were enhanced, to 8.8, 7.5, and 9.3 g/plant and 2.5, 2.0, and 2.6 g/plant, respectively following the treatment of three strains.

2.4   Microbes Improve Agricultural Production and Nutritional Quality

Several instances of harnessing the useful interactions between host plant and microorganisms explicitly indicate that agricultural yield could be enhanced considering nutritional quality and biofortification attributes (Rana et al. 2012; Anupama et al. 2015; Dubey et al. 2015, 2016b; Singh et al. 2018). Abiotic stresses are the key components in reducing crop production. However, the strength of these stresses changes with different soil and plant factors (Nadeem et al. 2014). Some adverse impacts of these stresses on the plants include nutritional and hormonal imbalance, and physiological and metabolic disorders such as epinasty, abscission, senescence, and susceptibility to the diseases (Nadeem et al. 2014). Most of the yield parameters and nutritional attributes are affected by stress conditions, which further affect numerous physiological and biochemical processes. Among these processes, N-fixation is one of the important processes that are susceptible to stress conditions. Apart from this, biosynthesis of metabolites such as amino acids, vitamins, phytohormones, and/or nutrient solubilization and mineralization processes are also prone to adverse conditions (Nadeem et al. 2014).

Research contextualizing to overcome the stresses and, hence, improve the production and the nutritional quality of the crop plants has been performed (Krey et al. 2013; Anupama et al. 2015). The yield of a wheat crop was enhanced with the inoculation of different PGPR such as cyanobacterial strains *Anabaena* sp., *Calothrix* sp., and *Anabaena* sp. and *Providencia* sp. (PW5) (Rana et al. 2012). Enhanced grain production with about 4.77, 4.87, and 4.66 t/ha with the combinations of different strains, that is, PW5, CW1 + PW5, and CW1 + CW2 + CW3, respectively. However, grain yield in absolute control was 4.18 t/ha. Further, the biomass yield was 15.33, 13.76, 13.56, and 14.94 t/ha with the respective combination of strains whereas the biomass yields in the absolute control were about 14.15 t/ha. Similarly, the harvest index was enhanced at 34.75%, 35.97%, and 31.20% by inoculation with the respective combinations. The nutritional component of wheat-like protein level and vital micronutrients such as copper, iron, zinc, and manganese was enhanced by inoculation of PW5 (Rana et al. 2012). An increase of 18.6% in protein level was observed in *Providencia* sp. (PW5) treatment supplemented with N60P60K60 in comparison to the control with N60P60K60 fertilizer only. The concentration of Fe was drastically enhanced in the wheat grains following inoculation of combinations of CW1, PW5, CW1 + PW5, or CW1 + CW2 + CW3, with the basal dose of N60P60K60. The concentration of Fe was 136.93, 271.93, 206.13, and 129.40 mg/kg, respectively; however, that in the control was 67.73 mg/kg. Similarly, concentration of Zn was 36.27, 41.73, 39.80, and 37.60 mg/kg, respectively, whereas the control was 31.60 mg/kg. The level of Mn was about 38.13, 53.40, 34.33, and 33.40 mg/kg, respectively, as compared to that of the control, which was 22.93 mg/kg. The level of Cu was 41.93, 99.00, 81.60, and 37.53 mg/ kg, respectively, while that of the control was 33.13 mg/kg (Rana et al. 2012). Biofortification of Zn in wheat was compared with the control when inoculated with three *Bacillus aryabhattai* strains MDSR7, MDSR11, and MDSR14 (Ramesh et al. 2014). The accumulation of Zn in the shoot was observed to be 19.7, 17.7, or 23.5 µg/g with the respective strains, which was higher in comparison to the control (14.0 µg/g). The accumulation of Zn was high in the root system in comparison to the shoot (25.1, 23.5, and 29.0 µg/g, respectively). Rice plants inoculated with *Paenibacillus kribbensis*, *Bacillus* spp., *Klebsiella pneumoniae*, and *Microbacterium* spp. showed increase in plant physical parameters including length, growth, dry biomass, and inhibition of fungal pathogens (Ji et al. 2014). Increased rice production has been recorded over the control upon the inoculation of rhizospheric diazotrophs in the semi-arid tropic grassland of India (Sarathambal et al. 2015). Apart from the N_2-fixing capability of diazotrophic isolates, synthesis and export of phytohormones may perform a key role in plant growth promotion and further increase the production. Furthermore, the study also suggested that the field treatment with *Serratia* sp. (CB2) and *Klebsiella pneumoniae* (CR3) enhanced the grain production by 31% and 28%, respectively, in comparison to fertilizer treatment with full doses.

In soybean, Zn accumulation was higher than the control (31.9 µg/g) in the plants inoculated with three *Bacillus aryabhattai* strains, MDSR7, MDSR11, and MDSR14 (39.2, 35.3, and 38.5 µg/g respectively) (Ramesh et al. 2014). There was a substantial increase in the nutritional quality (protein, prolamins, iron, and zinc) when plants were inoculated with the bacteria *Pseudomonas fluorescens* Pf4 and/or the arbuscular mycorrhizal fungi (AMF) fungi. Iron and Zn content (36.9 and 42.8 g/100 g of grain) was higher than the control (26.3 and 33.5 g/100 g of grain) in plants inoculated with only Pf4 strains. When inoculated singly with AMF, with both Pf4 and AMF, the iron and zinc content was higher, 33.9 and 42.6 g/100 g of grain and 34.0 and 39.5 g/100 g of grain, respectively. Maize plants showed higher protein content (10.1 and 10.0 g/100 g of grain) after its inoculation with AMF separately or together with Pf4 strain as compared to the control (9.5 g/100 g of the grain) (Berta et al. 2014).

2.5   Biocontrol Agents and Resistance Against Plant Diseases

The changing climate scenario and excessive mono-cropping practices have increased pest and disease incidences in crop fields (Dubey et al. 2016a). For managing diseases effectively, biocontrol agents could be a sustainable solution. In the line of this, *Simplicillium lamellicola* has shown effectiveness in controlling various diseases such as late blight, leaf rust, and powdery mildew in tomato, wheat, and barley, respectively. This biocontrol fungus produces antibacterial compounds such as halymecin F, halymecin G, and ($3R,5R$)3-O-β-D mannosyl-3-5 dihdroxydecanoic acid, commonly called mannosyl lipids, and a mycoparasite commonly called verlamelin (Dang et al. 2014). Furthermore, rice brown sheath rot disease caused by *Pseudomonas fuscovaginae* is controlled by the rhizobacteria *Bacillus amyloliquefaciens* Bk7, which has 93% pathogen suppression efficacy. *B. amyloliquefaciens*

Bk7 prominently shows the intrinsic mechanism of phosphate solubilization, IAA, siderophore, and ammonia production, and biofilm formation. Certain antimicrobial peptide genes such as srfAA, fenD, bmyB, bacA, and ituC showed higher expression in *B. amyloliquefaciens* Bk7 during pathogen exposure (Kakar et al. 2014). The mechanism of *Rhizoctonia solani*-mediated sheath blight disease of rice was explained by surfactin (srfA-A), the bacillomycin L (bacD) gene mutant, and srf + bac double mutants of *Bacillus subtilis* 916. In all the mutants, surfactin and bacillomycin L production decreases with reduced swarming motility, biofilm formation, and colonization of the rice sheaths. Such biofilm conserves the root exudates for the host plant health and protects it from the soil-borne pathogens (Beauregard et al. 2013). Mutant (srfA-A) is not able to restore biofilm formation with exogenous surfactin supplement (Zeriouh et al. 2014), whereas bacD mutant can form biofilm upon surfactin or bacillomycin L addition, indicating that the synergistic effect of surfactin and bacillomycin L is necessary for plant health (Luo et al. 2014). The lepidopeptide-mediated biocontrol via antibacterial activity can also activate induced systemic response in the host plant (Zaman and Toth 2013). Plant-associated microbes such as *Pseudomonas* sp. and *Chryseobacterium* sp. have the capacity to provide protection against biotic (*Xanthomonas campestris* pv. *oryzae*) and abiotic (salt; 3.5 g/l^{-1} in substrate) stress by activation of the induced systemic resistance (Lucas et al. 2014).

Microbial interactions with a wheat crop have an important role in disease suppression. *Pseudomonas fluorescens* HC1-07 suppresses the causal organisms of root rot and take-all of wheat, *Rhizoctonia solani* AG-8 and *Gaeumannomyces graminis* var. *tritici*, respectively. Disease suppression was regulated by the production of cyclic lipopeptides (CLP). Mutant studies reveal that the prtR and viscB genes are involved in the production of the viscosin-like CLP. The gene prtR has an additional trait of protease production that improves the defence system of the plants (Yang et al. 2014). *Pseudomonas fluorescens* further inhibits the growth of *Botrytis cinerea*, a phytopathogen in *Medicago truncatula* (Hernández-León et al. 2015). ACC-deaminase, siderophores, indole-3-acetic acid, phenazines, cyanogens, and proteases produced by *P. fluorescens* are signatory molecules for supporting PGP and plant health. *P. fluorescens* also has the ability to form A biofilm and secrete antifungal volatile organic compounds (VOCs), such as dimethyl disulfide, which has a proven role in plant defence (Huang et al. 2012). Thus, in any systematic approach, knowledge of soil microorganism biofilms, VOCs, and signalling in combination with plants is necessary for better crop health and improved agro-ecological services for sustainable agro-ecosystems (McGenity et al. 2018) (Table 2.1).

2.6 Microorganisms Improve Soil Quality

Soil is a living dynamic system that supports terrestrial life on Earth. There has been a steady and serious decline in soil health because of heavy use of agrochemicals, deforestation, and unregulated release of pollutants to the soil (Lehman et al. 2015).

Table 2.1 Major agricultural crops: their diseases and potential biocontrol agents against the diseases

Sample no.	Major group of crops	Crop name	Prevalent diseases	Causal organism	PGPM/biocontrol agent	Biocontrol efficiency (%)	References
1.	Cereals	Wheat	Powdery mildew	*Blumeria graminis* f. sp. *tritici* (Bgt)	*Bacillus subtilis* E1R-J	90.97	Gao et al. (2015)
			Take-all disease	*Gaeumannomyces graminis* var. *tritici* R3-111a-1	*Pseudomonas fluorescens* HC1-07	56.8	Yang et al. (2014)
			Rhizoctonia root rot	*Rhizoctonia solani* AG-8 strain	*Pseudomonas fluorescens* HC1-07	46.1	Yang et al. (2014)
			Leaf rust	*Puccinia recondita*	*Chromobacterium* sp. C61	–	Kim et al. (2014)
			Flag smut	*Urocystis tritici* Körn	*Bacillus thuringiensis* 58-2-1	94.0	Tao et al. (2014)
		Rice	Sheath blight	*Rhizoctonia solani*	*Bacillus subtilis* 916	66.7	Luo et al. (2014)
			Blight disease	*Xanthomonas campestris* pv. *oryzae*	*Pseudomonas* sp. BaC1-38	80.0	Lucas et al. (2014)
			Rice blast	*Magnaporthe oryzae*	*Chromobacterium* sp. C61	–	Kim et al. (2014)
			Sheath blight	*Rhizoctonia solani*	*Chromobacterium* sp. C61	–	Kim et al. (2014)
		Maize	Ear rot of maize	*Fusarium verticillioides*	*Trichoderma harzianum* T22	58.0	Ferrigo et al. (2014)
2.	Pulses	Chickpea	Southern blight	*Sclerotium rolfsii* Sacc.	*Trichoderma aureoviride* T42	60.0	Saxena et al. (2015)
		Bean	White mould	*Sclerotinia sclerotiorum* (Lib.) de Bary	*Trichoderma aureoviride* T42	82.2	Saxena et al. (2015)

(continued)

Table 2.1 (continued)

Sample no.	Major group of crops	Crop name	Prevalent diseases	Causal organism	PGPM/biocontrol agent	Biocontrol efficiency (%)	References
3.	Oilseeds	Soybean	Damping-off	Colletotrichum truncatum	Pseudomonas aeruginosa	97.0	Begum et al. (2010)
		Rapeseed	Leaf blight	Sclerotinia sclerotiorum	Fusarium oxysporum CanR-46, Aspergillus flavipes CanS-34A	94.5	Zhang et al. (2014)
			Grey mould	Botrytis cinerea	Fusarium oxysporum CanR-46	91.0	Zhang et al. (2014)
4.	Vegetables	Spinach	Wilt of spinach	Fusarium oxysporum f. sp. spinaciae	Bacillus amyloliquefaciens Q-426	86.3	Zhao et al. (2013)
		Tomato	Gray mould disease	Botrytis cinerea	Chromobacterium sp. C61	–	Kim et al. (2014)
			Late blight	Phytophthora infestans	Chromobacterium sp. C61	–	Kim et al. (2014)
		Chili	Anthracnose	Colletotrichum capsici (Sydow) Butler and Bisby	Trichoderma aureoviride (T42)	73.3	Saxena et al. (2015)
		Potato	Soft rot	Pectobacterium sp.	Paenibacillus dendritiformis	–	Lapidot et al. (2014)

Available nutrient levels are commonly low at unfertile/contaminated sites and resource competition, especially for nitrogen, phosphorus, potash, zinc, iron, etc., which become limiting factors for growth of the plants (de-Bashan et al. 2012). Microbial interactions with plant roots can improve soil quality in degraded, marginal, and unfertile land. Beneficial microbiota are a subset of the rhizospheric microorganisms that make the soil suppressive to soil-borne disease through production of iron chelators and antibiotics, controlled colonization of the root tissues, inter-species resource competition, immobilisation, and biodegradation of hazardous compounds, production of the signalling molecule salicylic acid, which induces systemic resistance in plants and enzymes such as ACC-carboxylase that degrades the ethylene precursor repressing the plant stress response to various stress factors (Nadeem et al. 2014; Lehman et al. 2015). The association of plants with arbuscular mycorrhizae also helps in improving soil quality as the produced glycoprotein "Glomalin" helps in soil aggregate formation. Plants also improve the soil quality by addition of readily available carbon inputs belowground as a distinct rhizo-secretion of the simpler form of compounds (mono-saccharides, low-carbon-containing amino and organic acids), mucilage, root debris (Philippot et al. 2013). Freshly added carbon could enhance the turnover of soil organic matter to increase nutrient availability (Trivedi et al. 2013), enhance fertility, provide the nutrient source to the soil microorganisms, and shape the soil microorganism community structure in the rhizosphere. Addition of root exudates at 32.75 mg kg^{-1} total organic carbon concentration was reported to enhance the xenobiotic (pyrene) degradation, shape the community structure of soil microorganisms, and enhance the activities of the enzymes catalase, dehydrogenase, and phosphatase in degraded soils (Xie et al. 2012; Sasse et al. 2018). Still, knowledge regarding how the different components of the root exudates shape community structure of soil microbes to promote plant health or restore degraded land is limited. Thus, we need to have an in-depth analysis of the molecular mechanisms of root exudation and plant–microorganisms interactions for successfully exploiting these to understand ecosystem services (Zhalnina et al. 2018).

2.7 Microbially Assisted Extensification for Improving Agricultural Production from Degraded Lands

Marginal lands are lands where cost-effective agricultural production is not possible (Edrisi and Abhilash 2016). Hence, beneficial plant–microbe interactions and soil biodiversity become critically significant in such lands to improve plant health (Tripathi et al. 2014a; Rillig et al. 2018). Experiments on the reclaimed desert lands significantly supported canola plant (*Brassica napus* L.) growth when amended with different microbial strains (El-Howeity and Asfour 2012). The amendments included *Azotobacter chroococcum*, *Azospirillum brasiliense*, and *Paenibacillus polymyxa*. Maximum yield was observed by the inoculation of *Azospirillum brasilense* supplemented with 60 kg N/fed as compared to other bacterial strains and the control treatments. Serw-4 variety of canola plant with 60 kg N/fed showed high seed yield/plant and seed yield/hectare (73.70 g and 4211.24 kg, respectively) compared to other tested varieties (El-Howeity and Asfour 2012). Various mechanisms are associated with lowering plant stress from different contaminants (Tripathi et al. 2015a, b), one of which may be the lowering of ethylene concentration under heavy metal stress which supports plant growth. Another mechanism can also be the accumulation of toxic metals by the microbial strains at the cellular level, hence reducing availability to the crop (Nadeem et al. 2014). Microorganisms such as arbuscular mycorrhizal (AM) fungi, viz. *Glomus mosseae*, promote maize plant growth under heavy metal-contaminated soils. The fungi produce an insoluble glycoprotein glomalin that acts as a chelating agent for heavy metals. Furthermore, the co-inoculation of AM fungi and certain bacterial strains such as *Brevibacillus* spp. promotes the growth of the red clover plant in contaminated soils (Nadeem et al. 2014).

2.8 Microorganisms for Phyto-Bioremediation, Carbon Sequestration, Biomass, and Bioenergy Production

The present scenarios indicate that food, energy, and emission of trace gases (CO_2, CH_4, N_2O, fluoride gases) are the key global challenges of the twenty-first century. The rapidly increasing population has resulted in accelerated consumption of fossil fuels, resulting in warming of the planet Earth and climate change (Dubey et al. 2016b; Edrisi and Abhilash 2016). In such a situation, renewable energy options from microbial sources (PGPR, PGPF, microalgae, cyanobacteria) could be green, sustainable, and carbon-neutral solutions (Ragauskas et al. 2006). Microbial biomass is a potent renewable energy source that has the capability to replace fossil fuels. Soil pollution by heavy metals, organic pollutants, and industrial effluents, as well as the energy crisis, are challenging issues. Besides the energy crisis, in India, approximately 1391.09 km^2, 58 km^2, and 593.65 km^2 land area are degraded by strong alkali, industrial contaminants, and mining pollutants, respectively (NRSC 2011). Thus, it is imperative to develop technologies for restoring degraded lands. Microbe-assisted phyto-remediation is a low-input sustainable technology in comparison to the conventional physicochemical and biological treatments and thermal desorption remediation methods. In degraded lands, rapid growth and high biomass/ bioenergy crops such as *Zea mays*, *Brassica napus*, and *Glycine max* associated with microbes may address the issue of increased global atmospheric CO_2 concentration by capturing it in the plant biomass (Witters et al. 2012). This method has a variety of benefits as it can capture carbon, hyper-accumulate heavy metals, remediate organic pollutants, and offer biofuel and raw materials for paper and pulp industry, as well as the revitalization of degraded land. Restored areas also balance increased demand for arable lands for energy and food production (Meers et al. 2010). Therefore, the practices of energy crop production can maximize agricultural land area by restoring contaminated lands with microbe-assisted phyto-remediation and sufficient food for the increasing human population. For better utilization of biomass produced through microbe-assisted phyto-remediation, it becomes necessary to address ecotoxicological studies of accumulated pollutants (Abhilash and Yunus 2011). Because the use of green carbon can slow the risk of atmospheric pollution from fossil fuel utilization (Tour et al. 2010), the annual crops of short-rotation woody biomass could be the best options to address the energy crisis issue (McKenney et al. 2011). Additional advantages of biomass may also be seen in increased soil fertility, microbial community enrichment, prevention of land erosion, and biodiversity maintenance (Abhilash et al. 2011; Tripathi et al. 2016a, b).

Biomasses produced by different microbe-assisted phyto-remediating plants have the capacity to revitalize and remediate the marginal or degraded lands by carbon sequestration, decreased greenhouse effects, and addition of organic carbon as rhizo-secretions and litterfall (Abhilash et al. 2013b; Zhalnina et al. 2018). Some plants have a fast N-uptake and N-use capacity that might concurrently improve the soil fertility (Gelfand et al. 2013). The majority of lands are contaminated with pesticides, fly ash, organic pollutants, and heavy metals. Restoration of these contaminated lands is possible through multipurpose phyto-remediation mediated by *Jatropha curcas*, *Vigna radiata*, and *Spinacia oleracea* (Tripathi et al. 2014b; Dubey et al. 2014; Edrisi et al. 2015). Plants such as *Eucalyptus*, *Pinus*, *Populus*, *Salix*, *Pongamia*, *Miscanthus*, *Camelina*, and *Panicum* have multipurpose benefits that could deepen the impact of biomass on land restoration, fertility, and food security (Graham-Rowe 2011; Tripathi et al. 2017). In addition to microbe-assisted phytoremediation, microorganisms help in biomass and bioenergy production (and also act as major players in terrestrial ecosystems). The associated processes of this system include carbon and nitrogen cycling, plant growth promotion, and greenhouse gases (CO_2, CH_4, and N_2O feedback system) (Harfouche et al. 2011). Alternative approaches such as transgenic technology for modifying plant root traits and harnessing benefits of plant–microbe interactions for remediation of the contaminated lands are suggested (Abhilash et al. 2012; Abhilash and Dubey 2015). These approaches can sequester more carbon along with enhanced biomass production that could be used further as a feedstock for bioenergy (Tables 2.2 and 2.3).

2.9 Major Challenges for Wide-Scale Utilization of Microbial Services

Plant–microbe interactions offer multiple benefits, but the process is underutilised and lacks wide-scale application. Although there are reports of the beneficial impact of plant–microbe processes on plant health, food production, and ecosystem services under controlled conditions, similar results are often not replicated during the field application. Most studies consider a single microbe or a combination of two or three microbes; however, in a natural system these microbes have to compete for niche adaptation in multipartite interaction with a large number of microbial populations (Hussa and Goodrich-Blair 2013). Various biotic and abiotic factors also affect interaction performance. Further, we have limited information regarding how these interactions vary with changing ecological parameters, time, and space

Table 2.2 Microbe-mediated sustainable phyto-bioremediation-generated options for biomass and bioenergy and carbon sequestration

Sample no.	Microbial species	Experimental plants	Target pollutants	Additional benefits	References
1	Cadmium-resistant autochthonous microbial consortium (AMC)	*Juncus maritimus*	Cadmium	Enhances phyto-remediation of salt marsh sediments contaminated with cadmium metal; increases metal phyto-stabilization capacity; promotes metal phyto-extraction	Teixeira et al. (2014)
2	Cr-tolerant rhizobacteria	*Helianthus annuus* L.	Chromium	Increases plant height, stem and head diameter, grain yield, oil content of seeds, and total biomass; increases phyto-extraction of metal	Bahadur et al. (2017)
3	*Pseudomonas* sp.	Grass species	Dichlorodiphenyl-trichloroethane (DDT)	Increases bioavailability of organic pollutants	Wang et al. (2017)
4	*Pseudomonas fluorescens*	*Glycine max* L.	Fenamiphos	With addition of SiO_2, increases degradation of fenamiphos	Romeh and Hendawi (2017)
5	*Pseudomonas libanensis* TR1, *Psychrobacter* sp. SRS8, *Bacillus* sp. SN9	*Brassica oxyrrhina*	Heavy metals, Ni accumulation	Increases plant growth, leaf relative water and pigment content under drought conditions; improves metal translocation	Ma et al. (2009), Ma et al. (2016)
6	*Pseudomonas* sp.	*Festuca arundinacea*	Oily sludge-contaminated	Increases shoot and root dry weight of fescue; enhances microbial activity and diversity in soil; improves degradation of TPH and PAHs	Liu et al. (2013)
7	Arbuscular mycorrhizal fungi (AMF)	*Avena sativa*	Petroleum	Improves soil quality; increases biomass; enhances soil microbial activity and phyto-remediation of saline-alkali soil contaminated by petroleum	Xun et al. (2015)
8	*Burkholderia fungorum* DBT1	*Populus deltoides*	Dibenzothiophene, phenanthrene, naphthalene, fluorene	Removes PAHs (up to 99%); increases root growth and biomass; alleviates agricultural non-point source pollution and landfill leachate phytoremediation	Andreolli et al. (2013)
9	*Pseudomonas koreensis* AGB-1	*Miscanthus sinensis*	Heavy metal soil	Increases biomass, chlorophyll, and protein content	Babu et al. (2014b)

#	Microorganism	Plant	Contaminant	Service	Reference
10	Plant growth-promoting (PGP) bacterial strains	*Cytisus striatus*	Diesel-contaminated land	Enhances germination, seedling vigour, and biomass production	Balseiro-Romero et al. (2017)
11	*Pseudomonas* sp UW3	Oats	Saline soil	Increases plant biomass; salt uptake; decreases soil salinity	Chang et al. (2014)
12	*Lasiodiplodia* sp. MXSF31	*Portulaca oleracea*	Heavy metals	Increases biosorption and bioaccumulation capacities	Deng et al. (2014)
13	*Pseudoxanthomonas*	*Festuca arundinacea* L.	Petroleum hydrocarbon	Increases degradation of high molecular weight (C21–C34) aliphatic hydrocarbons (AHs)	Hou et al. (2015)
14	*Bacillus cereus* ERBP	*Clitoria ternatea*	Formaldehyde	Enhances airborne formaldehyde removal	Khaksar et al. (2016)
15	*Pseudomonas putida* PD1	Grass species	Polycyclic aromatic hydrocarbons (PAHs)	Promotes root and shoot growth; protects plants against phytotoxic effects of phenanthrene	Khan et al. (2014)
16	*Candida* VITJzN04	*Saccharum officinarum*	Lindane	Enhances lindane degradation	Salam et al. (2017)
17	*Pseudomonas* sp. ITRH25	Carpet grass	Hydrocarbon	Enhances plant biomass and hydrocarbon degradation	Tara et al. (2014)
18	*Pholiota adiposa* SKU714	*Brassica napus* L.	Zinc	Enhances plant biomass, bioethanol, and lignocellulase production	Dhiman et al. (2016)
19	*Burkholderia* sp. GL12	*Zea mays*	Copper	Maximizes plant biomass and generates feedstock for bioenergy production	Sheng et al. (2012)
20	*Trichoderma virens* PDR-28	*Zea mays*	Heavy metals	Enhances biodiesel, and biomass for biogas production	Babu et al. (2014b)
21	*Escherichia coli* KO11	*Pennisetum purpureum*	Cadmium	Produces plant biomass as source for bioethanol production	Ko et al. (2017)
22	*Penicillium aculeatum* PDR-4	Sorghum-sudangrass	Arsenic	Promotes plant biomass for bioenergy production	Babu et al. (2014a)
23	*Trichoderma harzianum* T22	*Salix fragilis*	Heavy metals	Enhances wood quality and biomass for biofuel production	Adams et al. (2007)

Table 2.3 Microbially assisted bioremediation for the restoration of degraded land

Sample no.	Microbial species	Functional attributes	Experimental setup	Plant rhizosphere	References
1	*Pseudomonas aeruginosa*	Bacteria produces siderophores applicable in Cr and Pb accumulation in maize shoot; can be used for contaminated land reclamation	Pot experiment	*Zea mays* L. 'Benicia'	
2	*Sinorhizobium* sp. Pb002	Helps in Pb phytoextraction in combination with Triton X 100 and enhances plant biomass	Microcosms with 200 g soil (dw) per pot	*Brassica juncea*	
3	*Burkholderia* sp. D54	PGPR helps in Cs-contaminated land restoration	Pot experiment	*Phytolacca americana* Linn. and *Amaranthus cruentus* L.	
4	*Burkholderia cepacia*	Translocation of Cd and Zn to shoot	Under OTC, hydroponic experiment	*Sedum alfredii*	Braud et al. (2009)
5	*Azospirillum lipoferum* strains, *Azospirillum brasiliense* strain SR80	A PGPR used for crude oil remediation, removes 56.5% of contaminant (with 1% addition) over 14 days in malate-based media; promotes root growth of wheat	In vitro steel grid in 0.25 l chamber	*Triticum aestivum* L.	de Gregorio et al. (2006)
6	*Mesorhizobium huakuii* subsp. *rengei* B3	Helps in Cd remediation and has PGPR activity; *Arabidopsis thaliana* gene for phytochelatin synthase cloned in *Mesorhizobium* accumulated >35 nmol concentration of Cd^{2+}/mg	In vitro study	*Astragalus sinicus*	Sriprang et al. (2003)
7	*Kluyvera ascorbata* SUD165, 165/26, SUD165/26	Remediates Ni-, Pb-contaminated land, produces Zn-siderophore for plant biomass promotion activity	Pot experiments	Tomato (*Solanum lycopersicum*), canola, and Indian mustard (*Brassica*)	Burd et al. (2000)
8	*Rhizobium galegae*	Bioremediates benzene-, toluene-, and xylene-contaminated soil, enhances plant growth	In vitro and under growth chambers	*Galega orientalis*	Suominen et al. (2000)

No.	Organism	Description	Experiment	Medium/Plant	Reference
9	Pseudomonas putida, Enterobacter cloacae with Azospirillum brasilense	Removes PAH (average removal efficiency was twice that of land farming), 50% greater than bioremediation, 45% more than phytoremediation	Growth chamber experiments	Festuca arundinacea	Huang et al. (2004)
10	Pseudomonas aeruginosa KUCd1 (RS)	Used for wastewater treatment in Kolkata, has PGPR activity, can accumulate up to 8 mM Cd intracellularly	Pot experiment in growth chamber	Mustard and pumpkin	Sinha and Mukherjee (2008)
11	Enterobacter sp. NBRI K28, mutant NBRI K28 SD1(RS)	Revitalize fly ash-contaminated soils in Raebareli district, UP, India; promotes root biomass by 78%	Pot experiment in greenhouse	Brassica juncea	Kumar et al. (2008)
12	Pseudomonas sp. PsA4 and Bacillus sp. Ba32	Promotes plant growth, produces 1-aminocyclopropane-1-carboxylate (ACC) deaminase, siderophore, IAA, phosphate-solubilising chemicals used for remediation of Cr-contaminated soil near Chennai, India	Pot experiment in growth chamber	Brassica juncea	Rajkumar et al. (2006)
13	Phanerochaete chrysosporidium, Trametes versicolor, Coriolopsis polyzona	Degrades PCB to 25% of the total amount; shows more than 25% removal efficiency; has capability to degrade xenobiotics such as 2,4,6-trinitrotoluene, reductive dehalogenation of various chlorinated hydrogenation such as $CHCl_3$, CH_2CL_2, trichloroethane	In vitro experiment	Nitrogen-limited mineral medium (NMM)	Novotný et al. (1997) Khindaria et al. (1995)
14	Phanerochaete laevis HHB-1625	Degradation of various aromatic compounds such as PAHs; transformation of anthracene, phenanthrene, benzanthracene, and benzopyrene by extracellular production of ligninolytic enzymes	In vitro experiment	Nitrogen-limited liquid medium (NMM)	Bogan and Lamar (1996)

(Nadeem et al. 2014). We do have knowledge of the in vitro requirements of the microorganisms, but we still lack information about differential utilization of the nutrients by these microbes in the rhizosphere, phyllosphere, and endophytic compartments. Such information would be useful in enhancing the performance of particular microorganisms in their niches. The cost-effectiveness of the microbe-based products, response time, and their shelf life reduces the popularity of these products among users in comparison to synthetic chemicals. Still, successful utilization of plant–microbe interactions has great potential to protect plants, enhance soil fertility, and increase crop production, offering an alternative environment-friendly strategy for increasing crop production with reduced chemical inputs. We need a systematic strategy to increase our knowledge about the plant–microbe partnership with metabolomic, genomic, and transcriptomic approaches to utilize their potential effectively. The soil is being contaminated with various hazardous chemicals as it acts as a primary sink for pollutants. Because of such anthropogenic practices, agricultural lands are being degraded and are losing their productivity. Restoration of such contaminated, degraded lands becomes imperative to feed the burgeoning population. The changing climate could also alter tree speciation, root exudation, resource allocation in plants, and the fate of the soil pollutants (Abhilash et al. 2013b, 2015). Some microbes present in the soil have the capability to degrade the xenobiotic compounds, promote plant growth, and restore the degraded lands. It is important to decipher the vast diversity of microbial taxa present in the soil for the development of better soil management practices and to understand the shift in the microbial community under the changing climate regime (Abhilash et al. 2013b). A better understanding of microbial community structure will be helpful in determining the deleterious environmental impacts of climate change. Despite the importance of studying soil microbial ecology, very few populations of the soil microbial diversity have been captured to date and more than 99% of them are not culturable in existing laboratory setups (Singh et al. 2009). To overcome the problem of nonculturable microbiota, culture-independent approaches such as metagenomics can be used to explain the unknown microbial diversity from direct environmental DNA. Nowadays, myriads of methods are available to study the spatiotemporal structure of soil microorganism diversity. In the past decade, application of nucleic acid-based molecular technologies has completely revolutionized and added multidimensional orientation to studying soil microbial diversity. Now it is possible to capture the diversity of nonculturable communities in the soil by sequencing soil DNA using next-generation sequencing technologies. Acquisition of metagenomic data requires the organisation of short overlapping sequences into the full reading sequence and further analysing it as an expressible gene. Sequencing technologies that have helped to exploit metagenomics for structural and functional community analysis include 454-pyrosequencing, Illumina, and single-cell resolution. The single-cell resolution technique can provide a better idea about the sequence of microbial community inhabiting with low abundance in comparison to metagenomics whereas the advantage of meta-transcriptomics lies in the RNA-level knowledge ultimately covering gene expression and proteome-level activity about the soil microbiome (Segata et al. 2013).

Chapter 3
Methods for Exploring Soil Microbial Diversity

Abstract Belowground microbial processes are at the helm of terrestrial ecosystem functions, and the enormous diversity of soil microorganisms acts as a key player. Thus, understanding the community dynamics of microorganisms in the soil is essential to know their distribution, abundance, and structure. Further, it is also important to know how these communities are shaped in structure and function in response to changes in space and time. Various microbial diversity analysis methods—fluorescence in situ hybridization (FISH), denaturing gradient gel electrophoresis, terminal restriction fragment length polymorphisms, and the automated version of ribosomal intergenic spacer analysis, RISA (ARISA)—have been developed to analyse the diversity of soil microorganisms based on their genetic structure. However, methods such as phospholipid and fatty acid analysis utilise the differences in lipid components of the microbial cell membrane to analyse their diversity. In the present chapter, we explore the conventional methods of soil microbial diversity analysis.

Keywords Automated RISA (ribosomal intergenic spacer analysis (ARISA) ·
Community dynamics · Denaturing gradient gel electrophoresis (DGGE) ·
Fluorescence in situ hybridization (FISH) · Terminal restriction fragment length polymorphism (T-RFLP)

Since the inception of soil microbiology, various techniques have been employed for identification and characterization of soil microbial communities (Fig. 3.1). The microbial identification techniques commonly utilised for improving scientific understanding are explained in the further sections.

Fig. 3.1 Schematic representation of general workflow showing the three different stages for isolation, biochemical, structural, and functional characterization of soil microbial communities

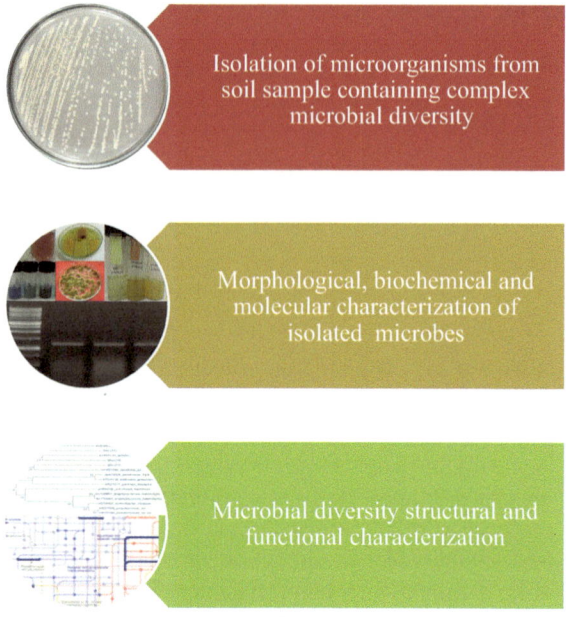

3.1   Phospholipid Fatty Acid Analysis (PLFA)

The fingerprinting technique does not relies on nucleic acid analysis; rather, it characterizes chemically diverse and abundant phospholipids as major constituents of the cell membrane to establish diversity in communities (Drenovsky et al. 2010). Some lipids (fatty acids) are the signature of specific groups of microorganisms (Zelles 1999; Ben-David et al. 2004; Kaur et al. 2005). The phospholipids consist of a polar head group and an ester-linked fatty acyl chain. The nature of the phospholipid varies in composition among eukaryotes and prokaryotes and even within different prokaryotic groups (Ben-David et al. 2011). Diversity in lipid composition across microorganisms and rapid degradation of phospholipids after their death make phospholipids suitable to analyse microbial diversity in nature (Zelles 1999). First, phospholipids are extracted from soil samples. The isolated phospholipids are then subjected to methanolysis for conversion to fatty acid methyl esters (FAME), which are explored by gas chromatography–mass spectrometry (GC-MS) for further identification. The total amount of phospholipid fatty acids obtained equals the viable amount of microbial biomass present in the soil. The different phospholipids obtained constitute different microbial groups, thus depicting the microbial community structure of the soils (Franzmann et al. 1996). PLFA analysis could also help in identifying the physiological stress in bacterial species (Kaur et al. 2005). This attribute of PLFA could be exploited while comparing the physiological status of the microbial community of different geographic locations (Wilkinson et al. 2002). In extreme arid conditions, phospholipids of the cellular remains of dead microorganisms also contribute to the phospholipid of viable microorganisms. Thus, results should be cautiously analysed when handling such soil samples (Lester et al. 2007). PLFA analysis is applied to study the vertical distribution of microbe communities and identify the environmental factors that contribute to the soil depth gradient (Li et al. 2017). The technique is also used to set biomarkers that show the beneficial response of soil microbial communities to soil management through crop rotations (Lupwayi et al. 2017). The difference in PLFA content provides an immense amount of information on the microbial community during sludge composting with an inorganic bulking agent (Wu et al. 2016). These data help in the structural-functional characterization of the microbe communities in contaminated soils such as semi-arid mine soil for assessing the mycorrhizal-amended phyto-management practices (Kohler et al. 2016). Various management practices are adopted in agricultural fields to improve soil organic matter (SOM), microbial functions, crop productivity, and agricultural sustainability. The structure of the microbial communities is specifically altered with each management strategy and these alterations can be easily indentified through PLFA analysis. For example, under rice field biochar amendment, soil type, cover crops, irrigation practices, and land use pattern affect the SOM quality and microbial community structure, crop rotation, and conservation (Calderon et al. 2016; Li et al. 2017). PLFA has been used in studying microbial community structure under soils contaminated by hydrocarbons, polyaromatic hydrocarbons, and heavy metals and acidic soils from mining (Markowicz et al. 2016; Zornoza et al. 2016).

3.2 Fluorescence In Situ Hybridization (FISH)

Microbial cells can be studied through applying the fluorescence in situ hybridization (FISH) technique without cultivating the microbes under in vitro conditions. FISH is used to detect small deletions as well as duplications that are not visible in microscopic analysis. The technique can be used to detect chromosome types and their rearrangements in microbial cells. FISH uses certain chemicals that fluoresce and are detected in the specific region on a chromosome (Teira et al. 2004). Through FISH, the microbial world in environmental samples (soil, water, contaminated land) can be characterized taxonomically by means of rRNA-mediated oligonucleotide probes and subsequently analysed for signal amplification through catalysed reporter deposition (CARD) (Ishii et al. 2004).

Microorganism cells are subjected to adhesion with chemical fixatives and hybridized with oligonucleotide probes on a glass slide or in a solution. The probes are generally 15 to 25 nucleotides in length and covalently labelled with fluorescent dye at the $5'$-C end. These probes are also linked with horseradish peroxidase (HRP) enzyme. As soon as the hybridization and the following deposition of fluorescent tryamides occurs, the signal on the target cells is intensified in comparison to the directly fluorochrome-labelled FISH probes. The FISH method has been used for analysing changes in bacterial populations of agricultural soils subjected to various pesticides and herbicides (Caracciolo et al. 2010). After proper washing, the distinctly stained cells are identified through flow cytometry or epifluorescence microscopy (Pernthaler et al. 2002). The suitable application of fluorophore-labelled rRNA-targeted probes as phylogenetic strains for cultivation-independent identification of microbes was first reported by DeLong et al. (1989) Since then, this technique has become a reliable and rapid method for identification of microorganisms in the environment (Sekar et al. 2003).

FISH is one of the commonly applied tools for localizing, identifying, and isolating desired microbial taxa in environmental microbiology. Single-cell techniques are promising for studying microbial community composition, and the efficiency can be further improved through FISH technology (Amann and Fuchs 2008). FISH identified the complex *Nitrospira* community which persisted in wastewater treatment plants, suggesting the stable coexistence of a *Nitrospira* strain that shared highly similar ecological niches with other microbial species (Gruber-Dorninger et al. 2015). Guangyin et al. (2016) analysed the electro-methanosynthesis process in a twin-chambered microbial electrolytic cell equipped with a graphite-covered hybrid biocathode through FISH. Microbial community structure of PAHs-contaminated river sediments was explored by using temperature gradient gel electrophoresis (TGGE) and a catalysed reported deposition (CARD-FISH) (Zoppini et al. 2016).

Fluorophore signal intensity is the most limiting factor in studying soil microbial communities using FISH technology. To overcome poor fluorescence problems in FISH, a recent advance is introduction of a single oligonucleotide combinatorial probe labelling called multi-labelled FISH (MiL-FISH). In MiL-FISH, the single-oligonucleotide probe combinatorial labelling is used to improve the signal intensity and image quality of individual microbial cells in environmental samples (Schimak et al. 2016).

3.3 Denaturing Gradient Gel Electrophoresis (DGGE)

Denaturing gradient gel electrophoresis (DGGE) is another popular 16S rRNA-based microbial fingerprinting technique that has been actively applied in exploring the microbial communities of natural ecosystems (Drigo et al. 2009; Janczyk et al. 2010). Amplification of the desired 16S rDNA fragment, mostly the V3 region and different amplicons, is followed by its further subjection to denaturing polyacrylamide gel electrophoresis (DGGE). Different amplicons have different electrophoretic mobility inside the gel, which leads to separation of the amplicons and generation of polymorphic bands in the gel. The community members can be identified by hybridization with taxon-specific probes. The differentiating bands could also be characterized by cloning and sequencing (Muyzer 1999; Jousset et al. 2010). DGGE has various advantages over other fingerprinting techniques as it offers rapid and intensive phylogenetic characterization and generates the structural profile of microbial communities. Alternatively, slight modification in DGGE such as associating it with other fingerprinting techniques could increase the resolution of the microbe communities (Jousset et al. 2010; Costa et al. 2006). DGGE has certain limitations: similar to any other 16S rRNA-based technique, it also faces the artifacts of polymerase chain reaction (PCR)-based amplification (Martin-Laurent et al. 2001). An inherent problem with DGGE is that all the samples cannot be loaded on one gel; thus, variations among the different gels also produce lower reproducibility (Nunan et al. 2005). TGGE is another variant of this technique, in which temperature acts as the denaturing agent instead of chemical-based denaturation of the DGGE (Ogier et al. 2002). DGGE helped to identify potential aluminum-tolerant bacterial strains, which improved plant health and phosphorus nutrients in ryegrass cultivated with cattle dung manure in volcanic soil (de la Luz Mora et al. 2017). The technique helped to map diverse microbial community profiling under the heavy metal-polluted soil (Becker et al. 2006). Application of DGGE along with Biolog Ecoplates and microbial biomass helped to analyse the impact of glyphosate on microbe-mediated soil properties, which deciphered glyphosate-mediated structural-functional changes in communities of microbes after 15 days of herbicide application (Mijangos et al. 2009). In *Zea mays*, DGGE showed the protease activity of the rhizosphere and bulk soil microbes with different N uptake, leading to the conclusion that the abundance of *npr* and *apr* genes increases nitrogen use efficiency along with the higher soil enzyme activity (Baraniya et al. 2016).

3.4   Terminal Restriction Fragment Length Polymorphism (T-RFLP)

Terminal restriction fragment length polymorphism (T-RFLP) has been widely used for exploring the predominant microbe population in different habitats. It also helps in studying the spatiotemporal changes in the community structure of microorganisms (Lukow et al. 2000). T-RFLP is a high-throughput fingerprinting technique exploiting the 16S rRNA sequences of prokaryotic organisms. However, 18S rRNA genes are used for fungal diversity analysis (Liu et al. 2016a, b). These rRNA sequences are amplified from the soil DNA samples using fluorescent primers. Further amplified sequences are then cleaved with the desired restriction enzyme (RE) having four base-pair restriction sites; the RE with four base-pair restriction sites has higher frequencies of these sequences in the DNA samples. The number and the size of the fluorescent terminal restriction fragments (TRFs) are then analysed through a DNA sequencer. The diversity among the phylogenetically distinct microorganisms is reflected by the differences in the length of TRFs. Thus, the structure of the numerically dominant communities of the microbes can be analysed by T-RFLP. As in a PCR-dependent technique such as T-RFLP, primer selection should be specific to the targeted microbial group and also generalized in nature, so that most of the desired bacterial population can be amplified. Sometimes, different microbial populations may have similar TRF size for a primer–RE combination. In such situations, use of more than one RE gives a better prediction of the microbial communities because TRF size for one primer enzyme combination could be similar but not for the other combinations. T-RFLP is extensively used in the studies of structural attributes of the microbial communities under an agricultural system (Malghani et al. 2016; Liu et al. 2016a, b) and also for the restoration of contaminated sites and bioenergy feedstock production (Mossa et al. 2017).

3.5 Automated Version of RISA (Ribosomal Intergenic Spacer Analysis (ARISA)

The automated version of ribosomal intergenic spacer analysis (RISA), ARISA, is a useful tool for exploring microbial diversity (Lee et al. 2012). It is a sensitive and highly reproducible technique in comparison to RISA and exploits the length heterogeneity of the internal transcribed spacer (ITS) that lies between 16S and 23S rRNA sequences of the ribosomal RNA operon. PCR amplification of the ITS sequence is done with 5′-fluorescently labelled primer. After being labeled, PCR products are analysed by an automated DNA analyser, and the results show polymorphism in the PCR product length. The length heterogeneity of each PCR product represents an operational taxonomic unit (OTU). Cloning and sequencing of the ARISA-PCR products and their further BLAST analysis reveals the identity of the microbes. Another approach for rapid phylogenetic characterization of microbial communities could be the direct phylogenetic analysis of the T-RFLP products through genetic libraries (Brown and Fuhrman 2005; Collins et al. 2006). ARISA offers more phylogenetic resolution as it targets the 16S-23S ITS rather than the 16S rRNA region targeted in T-RFLP studies (Hewson and Fuhrman 2004). Some microorganisms share ITSs of equal length and their diversity would not be shown by ARISA. On the other hand, some microorganisms may have multiple copies of rRNA operons having ITS of variable regions. Thus, a single microorganism may be overrepresented as different microorganisms in a community profile (Collins et al. 2006). Because ARISA is a PCR-dependent technique, it may face some artefacts of PCR-based amplification, but it offers better resolution of bacterial communities at species and strain levels (Kovacs et al. 2010a, b). The technique is economic and relevant for microbial community structure analysis under conservation agriculture practices (Likar et al. 2017; Butterly et al. 2016; Bertani et al. 2016) and for contaminated land phytoremediation studies (Fernandes et al. 2017).

3.6 Single-Strand Conformation Polymorphism (SSCP)

Single-strand conformation polymorphism (SSCP) is a screening technique for identifying different genomic variants in a microbial population from environmental samples (Fischer and Lerman 1979). The technique can detect sequence variation means, single-point mutations, or smaller changes in nucleotides, on the basis of differences in electrophoretic mobility. Mutated DNA bands have different mobility rates in comparison to wild-type DNA (Hayashi 1991). SSCP is a pivotal technique for evaluating microbe ecology and diversity in the soil system. The dissimilar electrophoretic mobilities and conformational differences are separated through nondenaturing polyacrylamide gel electrophoresis (Widjojoatmodjo et al. 1995).

SSCP is comparatively simpler in approach than DGGE or TGGE as it does not depend on denaturing gradient gels, GC clamp primers, or a specific apparatus. However, the technique is most suitable for 150- and 400-bp DNA fragments (Muyzer 1999). Yet, the increased frequency of DNA strands reannealing after denaturation in cases of higher DNA concentration is a limiting factor in deciphering high-diversity communities (Selvakumar et al. 1997). Again the problem lies with the presence of more than one band from the double-stranded DNA-based PCR product after electrophoresis, as similarly confirmation of complementary single-stranded products of double-stranded DNA may lead to the detection of less than three products from a single organism (Lee et al. 1996). The poor detection limit of the technique can be minimised using a fragment (~400 bp) of the microbial 16S rRNA gene (V4 and V5 sequence regions), which is PCR amplified with universal primers with the 5′-end of a primer phosphorylated. However, despite certain limitations, the technique has been effectively used for microbial community analysis in various habitats. SSCP was successfully applied to differentiate between interspacer regions of 16S–23S rRNA of bacterial strains (Scheinert et al. 1996). Further SSCP has been employed to distinguish the soil microorganisms *Pseudomonas fluorescens, Bacillus subtilis*, and *Sinorhizobium meliloti* (Schwieger and Tebbe 1998). SSCP-based exploration of the *Medicago sativa* and *Chenopodium album* rhizospheres revealed that bacterial communities are shaped according to plant species despite both plants growing in the same soil because rhizospheric bacterial communities varied for each plant.

The technique further revealed the diversity and activity of bacterial biofilm communities growing on hexachlorocyclohexane (HCH)-contaminated soil polluted by the pesticide-producing factories residues and waste material in Egypt (Gebreil and Abraham 2016). Moreover, the effectiveness of tomato-linked rhizospheric bacteria applied in single and consortium mode for control of tomato stem rot by *Sclerotinia* (Abdeljalil et al. 2016) was assessed using SSCP. Similarly, under arsenic (As)-polluted soils, 16S rDNA-based capillary electrophoresis single-strand conformation polymorphism confirms the altered soil microorganism community in response to concentration gradients of As (Quemeneur et al. 2016).

3.7　Stable Isotope Probing (SIP)

Stable isotope probing (SIP) is a common and suitable tool to establish appropriate connection between the structure and function of microorganism communities when it is linked with the metagenomics (Chen and Murrell 2010). Using this technique we can identify the microbial community in an environmental sample (Chen and Murrell 2010; Dumont and Murrell 2005) with biomarkers such as DNA and RNA. The technique is based on the labelling of growth substrate with stable isotopes such as ^{13}C, which is further incorporated in the cells of microorganisms as phospholipids, DNA, or RNA. Further, these SIP-tagged nucleic acids and phospholipids can be easily identified against the unlabelled background by using density-dependent centrifugation. DNA and rRNA are the biomarkers that give an idea about microbial diversity. DNA-SIP, PLFA-SIP, and RNA-SIP are the three major techniques reported so far. DNA-SIP involves amplifying the isolated DNA through PCR, then sequencing the 16S rRNA gene and further analysing the phylogeny to decipher functions of microorganisms along with their community structure (Dumont and Murrell 2005). The other type of SIP, called PLFA-SIP, has also been used in the study of microbial identity along with metabolic profiling in the environmental sample (Chen et al. 2008). The sensitivity of PLFA-SIP is much higher than the DNA- or RNA-mediated SIP (Neufeld et al. 2007). On the basis of PLFA-SIP profile, it is possible to mark most of the diversity inhabiting the habitat (Taylor et al. 2013). RNA-SIP is a beneficial technique to understand the ongoing microbe-based processes in the environmental samples. The procedure involves isolation of total RNA followed by a density-dependent centrifugation with 16S rRNA RT-PCR amplification. Further, DGGE profiling of the PCR products gives a detailed picture of major microbial communities involved in environmental processes (Manefield et al. 2002). RNA-SIP is more responsive than DNA-SIP but has major limitations related to the instability of the mRNA.

3.8　Quantitative PCR (Q-PCR)

Quantitative real-time PCR (qPCR) has been used extensively as a tool for identifying microorganisms of interest and their gene expression from different environmental samples (Higuchi et al. 1992). Real-time PCR provides appropriate quantification of targeted nucleic acids from even a low amount of the complex starting material. Moreover, Q-PCR is a versatile tool for highly precise, extremely responsive, and high-throughput identification and quantification of targeted nucleic acid sequence from samples of diverse environmental compartments (Sanzani et al. 2014).

Application of Q-PCR strengthened our understanding about the abundance and role of microbial diversity under various ecosystems. Studies showed a direct link between the abundance of the specific compositional soil bacterial community and their denitrifying gene abundance genes to the greenhouse gas (nitrous oxide) emission from agriculture lands (Morales et al. 2010). The microbial community composition of Crenarchaeota in terrestrial as well as in aquatic ecosystems is also identified through Q-PCR, which confirmed that Crenarchaeota is a stable community in the terrestrial ecosystem (Ochsenreiter et al. 2003). The relative abundance of common groups of soil microorganisms was identified through taxa-specific quantitative PCR (Fierer et al. 2005). However, in some cases the Q-PCR base analysed abundance of microorganism groups not truly representing their percentage in environmental samples because of poor DNA isolation, improper rRNA sequence, and a heterogeneous number of rRNA operons (Smith and Osborn 2009; Fierer et al. 2005). Quantitative genotyping and single-nucleotide polymorphism (SNP) detection, identification of alleles, can easily be performed through Q-PCR. It has further implication in molecular diagnostics for rapid identification and enumeration of pathogens, which helps in controlling disease outbreaks in agriculture as well as in medical sciences. Q-PCR has rapidly gained popularity because of its robustness and higher responsiveness even for a low quantity of nucleic acids from different samples (Smith and Osborn 2009). Some researchers such as Smets et al. (2016), Props et al. (2017), and Wang et al. (2018) suggested that the Q-PCR can also be utilized for identification and quantification of gene expression, analysing the microbial abundance profiles in the soil and other complex environmental samples. Q-PCR-dependent detection tools based on species-specific primers are highly useful for providing insight about host–microbe interactions in different domains of the environment dealing with soil microbiology, ecology, etiology, and epidemiology of plant-pathogenic microorganisms.

3.9   DNA Microarray

Molecular-based methodologies have greatly increased our capability of microbial identification from different environmental compartments. Still, it is difficult to completely explore the vast microbial diversity of different environmental compartments as most of the methods can study only a small sample size with a limited set of organisms. We therefore need more comprehensive and robust methods. Microarrays with specifically designed microarray plates have unprecedented ability to quantify the microbial communities in a system. The method has been customized to access differences in gene expression between many cells and even in the same cell grown in dissimilar conditions (Zhou and Thompson 2002; Bodrossy and Sessitsch 2004; Schadt and Zhou 2005). DNA microarrays consist of probes composed of nucleic acids associated to a planar glass surface that is usually labelled with chemically reactive groups (epoxy, poly-L-lysine, or aldehyde) for proper binding of nucleotide probes. Samples are labelled chemically or by enzymatic reaction to analyse the presence of targeted nucleic acid sequences. Further labelled samples are hybridized on the array and washed with different-strength buffer solutions. The signal obtained through specific interactions among the probes and target nucleic acids is evaluated by a confocal microarray scanner. The probes hybridized to a labelled target will only yield a signal revealing the presence of cognate nucleic acid motifs in the sample (Huyghe et al. 2009).

For the purpose of microbial ecology, the microarray with specific variations is being utilised to study structural and functional characteristics of microbial communities. 16S rRNA gene microarrays are designed to provide the structural information whereas functional gene arrays (FGA) are formulated for explaining the functional attributes of the microbial communities (McGrath et al. 2010). An interesting study applied the GeoChip 3.0 microarray technology that proved significant variation in nitrogen-, carbon-, sulfur-, and phosphorus-cycling genes and processes associated with soil microbial communities under traditional organically managed agricultural fields (Xue et al. 2013). Microarray is also helpful in the studies of associated phytopathogens. Detection of pathogenic bacteria in potato causing blackleg and soft rot (*Pectobacterium atrosepticum, Dickeya* spp.), ring rot (*Clavibacter michiganensis*), scab (*Streptomyces scabies, Streptomyces turgidiscabies*), and brown rot (*Ralstonia solanacearum*) were identified via microarray (Degefu et al. 2016). However, cross-hybridization is the key limitation in microarray when analyzing environmental samples.

Chapter 4
Single-Cell Genomics and Metagenomics for Microbial Diversity Analysis

Abstract Soil metagenomic analysis was previously limited by technological restrictions and the few reference genomes. The advent of next-generation 'omics' technologies has provided high-throughput methods for analysing community structure and reconstructing soil metagenomes. High-throughput sequencing technology and single-cell genomics have revolutionized metagenomic analysis by enabling large-scale sequencing at reduced sequencing costs with less time required. In the present chapter we discuss various technological advances in metagenomics, their processes and the methods of data analysis, and metagenomic success stories under various environments that can be applied for studying the functional and structural diversity of soil microorganisms.

Keywords Functional annotation · Microbial community structure · Next-generation sequencing (NGS) technology · Single-cell genomics · Metagenome

Polymerase chain reaction (PCR)-based microbial diversity analysis has its own limitations. The isolation process of DNA and RNA bases in PCR results and harsh isolation methods cause shearing of the DNA, which creates problems in PCR detection and primer annealing. Environmental samples have a large amount of humic acid that usually is coprecipitated with DNA and induces bias during molecular microbial diversity analysis. Sequences 16S and 18S rRNA, and internal transcribed spacers (ITS), are highly conserved regions among all organisms. Differential amplification of these genes biases the results of PCR-based prediction of the microbial community. Several microbial genomes have high guanine-cytosine (GC) content, a different copy number of template gene primer specificity, and a high rate of hybridization in the soils (von Wintzingerode et al. 1997), and these differences cause potential bias in the PCR-based community structure prediction. Advances in the identification of microbial diversity are still ongoing, and chip-based methods have recently been under investigation (Stanley and van der Heijden 2017).

4.1   Single-Cell Genomics (SCG)

Most of our present knowledge about the genome and its control is gained from population-level studies involving thousands to millions of cells on average for analysis. The resulting study, even though informative, quite frequently ignores any heterogeneity within the population of cells (Macaulay and Voet 2014). Molecular differences among individual single cells inside tissues and organ systems have led to greater understanding. What is the dissimilarity between adjacent cells? How do neighbouring cells affect each other? How do the cells influence the organization and function of the organs and organisms? What is the difference at the genetic, epigenetic, and gene expression levels? What is the phenomenon behind specific development or diseases in individual cells? All these questions have remained unexplored (Lovett 2013). Now, the 'omics' of single cells offers the chance for exploring all these biological complexities that are currently a complex and challenging task (Lovett 2013). Development of the field of single-cell genomics signifies a turning point in cell biology. Through single-cell genomics, the expression level of each gene per genome across thousands of individual cells can be accomplished within a single experiment. Single-cell genomics can assist the discovery of new genes and pathways that regulate cell fate and transitions (Trapnell 2015).

Single-cell genomics and metagenomics correspond to two different sides of the same coin, providing an understanding of the biology of organisms that cannot be cultured (Perkel 2012). Metagenomics explores the genetic potential of a community but not the individuals. Single-cell approaches can cross that gap by endowing scaffolds for the assembly of metagenomics data or reference genomes for variation analyses. Therefore, for a large group of researchers these two approaches are complementary (Perkel 2012). The shift from 16S rRNA gene diversity analysis to metagenomics, and just recently to single-cell genomics (SCG), is largely driven by recent technological advances in sequencing approaches facilitating megabase-scale to, ultimately, terabase-scale surveys (Woyke and Jarett 2015). In cultivation-independent genomics approaches, SCG can exclusively facilitate assessment of the functional potential of totally unknown microbes without needing to rely on precise assembly and binning approaches for metagenome data (Woyke and Jarett 2015). One of the prime advantages of SCG is that it allows examination of samples at the basic unit of life, the individual cells, and permits access to nucleic acids and the relationship of features such as plasmids and cell-contained phages to the genome. Single-cell genomes are now consistently sequenced and submitted to public databases, similar to metagenomes 10 years ago (Woyke and Jarett 2015). SCG has been used in deciphering microbial entities and their communication behaviour in studying the interactions between two genes (Blainey 2012). The technique can provide the structure of the microbial community and their functionality in various ecosystems (Lasken 2012). It can also give in-depth knowledge about inter-organism metabolic interaction and the evolutionary traits of uncultivable microbes (Stepanauskas 2012). Single-cell genomics uses unique synergy between the complex instrumentation, in which fluorescent in situ hybridization/fluorescence-activated cell sorting (FISH/FACS) was used for separation of cells during the study of environmental samples such as soils (Abulencia et al. 2006). This flow cytometry is connecting the hub technology necessary for the exploration of the heterogeneous microbial population (Muller et al. 2010). The recent whole-genome amplification (WGA) process is boosted through nanoliter multiple-displacement amplification (Marcy et al. 2007), which results in accurate amplification without any biases. The resulting amplicons are perfect for consequent genome assembly, production of a sequence library, and generation of the full genome using high-throughput pyrosequencing. Reports showed that the excellence of the cell sorter and WGA unravelled the mystery of the uncultivated microbial world (Walker and Parkhill 2008) and minimized the drawbacks of PCR-based analysis of uncultivable microbial diversity. Technological advancements in the area of single-cell isolation, complete genome or transcriptome sequencing, specifically next-generation sequencing (NGS), and genome-wide analysis platforms through high-end computational approaches opened new avenues for high-resolution analysis of single-cell genomes or transcriptomes to disclose the hidden biological complexity (Macaulay and Voet 2014). Next-generation sequencing is significant in facilitating an extremely sensitive analysis of gene expression, epigenetic configuration, nuclear structure, and other features of the cellular state (Trapnell 2015). Numerous sequencing assays were optimized at the level of individual cells in the past few years and modifications stemmed from advances in instruments that physically capture and isolate individual cells and from improvement in amplification technique, and reverse transcriptase, which enhances the process (Trapnell 2015).

Sequencing single bacterial cells was not feasible until very recently because most of them contain only a minuscule amount of genetic material that is difficult to extract and process (Amann and Fuchs 2008; Lasken 2012; Dhillon and Li 2015). Shotgun-sequencing methods of DNA isolated from soil samples were the first breakthrough that facilitated ease of access to sampling environments (Lasken 2012; Dhillon and Li 2015). The second most important strategy was the use of multiple displacement amplification, which facilitated precise amplification of isolated genome sets, to ultimately reconstruct the genome of interest using bioinformatics algorithms (Lasken 2012; Kind et al. 2013; Lecault et al. 2012; Dhillon and Li 2015). Single-cell amplification was developed through the ɸ-29 (phi 29) DNA polymerase-dependent replication of mini-circular-deoxyribonucleic acid templates (Lasken 2012; Martínez-García et al. 2014; Dhillon and Li 2015). Initial efforts on reducing amplification bias were under-represented but with constant modification in the protocols, in silico corrections, and revalidation with the authentic repository of databases, minimum biases were found (Marcy et al. 2007; Lasken 2012; Martínez-García et al. 2014). Still, the process of targeted genome assembly from the amplified fragments is challenging (Rinke et al. 2013; Dhillon and Li 2015). In maximum metagenomic studies of a single-cell genome, assembly of the genome of a particular species apart from the most abundant species is an extremely difficult task (Marcy et al. 2007; Macaulay and Voet 2014; Dhillon and Li 2015). Precisely created assemblies from these samples were generated as a consensus genome from numerous fragments (Kamke et al. 2012; Dhillon and Li 2015). To obtain the best quality single-cell data and ensure separation of technical noise from the signal,

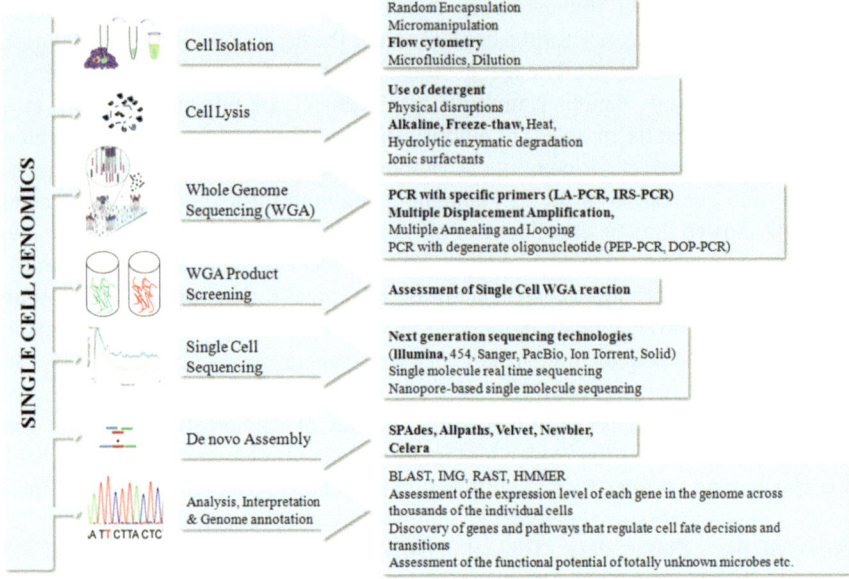

Fig. 4.1 Step-by-step analysis of single-cell genomics (SCG) elaborates the understanding of the microorganisms physiology, metabolisms, evolutionary linkage, underlying pathways, and their inter- and intra-species interactions under various ecosystems

every step needs cautious inspection when performing single-cell experiments (Gawad et al. 2016; Kalisky and Quake 2011). A brief description of the different steps involved in single-cell genomics is illustrated in Fig. 4.1.

4.1.1   Cell Isolation

The initial step in the isolation of particular cells from soil samples is to create a live single-cell suspension. This process cannot be neglected while working with a soil sample that contains various forms of tissues, which require a series of mechanical or enzymatic treatments to produce a viable cell suspension (Emmert-Buck et al. 1996). Soil microbial samples from various environments also need precise lysis of bacteria with prerequisites that can be highly inconsistent among different microbial strains (Zhou et al. 1996; Gawad et al. 2016). In suspension, various methods were evolved for isolation of single cells including manual manipulation such as serial dilution (Ham 1965), microwell dilution (Gole et al. 2013), micropipetting (Zong et al. 2012), and optical tweezers (Landry et al. 2013). Furthermore, myriads of protocols were standardized to separate intact cells by using the fluorescence-activated cell sorter (FACS) system (Navin et al. 2011). Nuclear isolation can strengthen the benefit of allowing single-cell analysis from a frozen tissue sample (Leung et al. 2015; Gawad et al. 2016). Sample preparation and FACS optimization are tedious when handling microbial samples from various sources (Rinke et al. 2014; Gawad et al. 2016). Automated micromanipulation approaches consisting of micromechanical valves in microfluidic devices are coming to the forefront of application under single-cell genomics (White et al. 2011; Leung et al. 2012; Macosko et al. 2015; Gawad et al. 2016).

Whatever the method applied, it is essential to check that an individual cell has been isolated properly to avoid a false interpretation. In an ideal case, this can be achieved through microscopic data of each single cell per well. Furthermore, identification of scalable approaches for separating single cells demands active research that may lead to the development of precise tools for all capture performance metrics (Gawad et al. 2016).

4.1.2   Whole-Genome Amplification

For obtaining significant genetic information from single cells, amplification of a single copy of the genome along with reducing different artifacts such as amplification biases, genome loss, mutations, and chimeras are essentially required. In one approach, the complete genome from targeted single cells is equipped with PCR-based amplification by using a universal sequence throughout the genome (Lichter et al. 1990), a universal sequence ligated to sheared genome (Troutt et al. 1992), or degenerate or random oligonucleotide priming (Telenius et al. 1992; Zhang et al. 1992). The second strategy involves methods based on isothermal approaches and multiple-displacement amplification (Zhang et al. 2001). Furthermore, to remove the poor resolution of PCR approaches and isothermal biases, new methods involve restricted isothermal amplification then subjected to PCR of previously amplified products (Zong et al. 2012). These combined approaches are at the forefront of the WGA methods in existing single-cell studies (Gawad et al. 2016).

4.1.3   Interrogation of WGA Products

Application of a single technique under single-cell study is not sufficient to produce whole-genome amplicons and copy number estimation. The kind of genomic interrogation opted for the study of single-cell genomics is chosen after considering the deviation of whole-genome amplification (Gawad et al. 2016; De Bourcy et al. 2014). Study of a location-specific genome within a single cell may assist us to find the physiologically active area that confers concise and coherent results and to save the cost and time of adopted techniques. Small and fewer variant genomes have minimum chances of including the technical deviations and errors that generally occur during the initial cycles of complete genome amplification. These technical deviations lead to incorrect results that often produce erroneous positive signals

(De Bourcy et al. 2014; Gawad et al. 2016). Whole-genome analysis of a single cell or an entire sample is a prominent methodology helping in identification of variations in single nucleotide and copy number (SNVs and CNVs) via reducing the exome-related errors. Furthermore, WGS can explore the intronic region and insertion and deletion of translocated gene segments actively involved in most biological systems (Gawad et al. 2016).

4.1.4   Overview of Single-Cell Sequencing Errors

Single-cell sequencing is a highly precise technique having less chance of errors whereas sometimes errors happen during isolation of a single cell from the complex tissues of the bulk sample, genome amplification, and copy number estimation. Characteristics of the cells such as size and phases of cell division also mark the cause of deviations in the final output (Hou et al. 2012). Single-cell sequencing of a microbial cell can be done only when precise and sensitive techniques for cell disruption followed by combined and efficient amplification and copy number estimation strategies are considered for the study (Zong et al. 2012; Deleye et al. 2017). Apart from the in-depth picture of single-cell phases, single-cell genomics also provides detailed information to users about the cell transition phases, although the field is still new and requires various experimental and computational advances for complete exploration of the potential (Trapnell 2015). Single-cell genomics studies need new algorithms and software for identification of differentially expressed genes. Continuous development of single molecule-based strategies via single-molecule real-time sequencing and nanopore-mediated single-molecule sequence analysis may avoid the need to amplify an individual cell genome and ultimately allow direct sequencing of the DNA present in an individual environmental cell, including its epigenome, representing a major breakthrough in this area (Woyke and Jarett 2015).

4.2   Metagenomics

Each microbe in any environment has its specific set of gene pool and genome. Metagenomics study covers the entire genome of all microbes inhabiting any habitat including soil and water without in vitro culturing, prior individual identification, or amplification (Abulencia et al. 2006; Kunin et al. 2008). Recently, it has been used as a prominent tool for the analysis of interacting soil microbes and their specific and interlinking functions. Remarkable developments were observed in metagenome sequencing using NGS technologies, which have generated enormous amounts of data. The technique comprises isolation of metagenomics DNA directly from environmental sample, fragmentation, generation of sequence clone library, and high-throughput sequencing to acquire detailed information. Function- and sequence-driven screening can lead to functionality of the metagenome. Later on, sequencing technologies such as Roche 454, Illumina, and Applied Biosystem SOLiD™ rendered proper understanding of complete genome sequencing without using traditional PCR or cloning. Because almost 99% of the microbes in various environments are still far being cultured in media, metagenomics offers a path to identify the complete microbiome profile, phylogenetic relationship, species diversity and abundance, metabolic abilities, and functional characteristics of the inhabiting microbes (Shah et al. 2011). Metagenomic approaches have emerged as a hub for understanding the ecological and evolutionary record of microorganisms, which is perhaps the most important, vital, and less explored biological area because of the diverse millions of metagenomic reads and their functional implications (Gilbert et al. 2011; Zarraonaindia et al. 2013) (Fig. 4.2).

NGS has facilitated high-speed generation of huge sequence data that can decipher the real picture of soil microbes in water, root rhizosphere, rhizoplane, and phylloplane, and the body of humans, especially the gut, animals, fishes, insects, and other organisms (Shokrall et al. 2012; Bai et al. 2014; Hanning and Diaz-Sanchez 2015). Metagenomic data signify functional attributes of complex belowground microbes, their intra- and interactions and ecological services and thus help to understand ecological and evolutionary aspects of the microbial ecosystems as genetic and metabolic networks (Filippo et al. 2012; Ponomarova and Patil 2015). Complex metagenomes can reflect how ecosystems are functioning and how biotic and abiotic changes in the environment influence whole community functions in the ecosystem. Thus, sequence assembly, annotation, analysis, phylogenetic surveys from the metagenome data, gene-centric approaches, and functional characterization of microbial communities have remained major challenges for metagenomic data analysis using bioinformatics protocols, pipelines, and data handling tools (Thomas et al. 2012; Wooley and Ye 2009).

4.2.1 Metagenomics and Soil Microbial Diversity

Soil contains $10^8 - 10^{10}$ microbial cells per gram (Raynaud and Nunan 2014) that accomplish life-sustaining functions for the planet. Soil is a biological entity that remains largely unexplored although we know a little about its rich diversity of microbial life (Liu et al. 2016a). Soil contains billions to trillions of microbial cells per gram, including diverse species of bacteria, fungus, archaea, viruses, and unicellular protists (Jansson 2011) that are involved in nutrient turnover and improve plant biomass and productivity (Prakash et al. 2014). Suppressive soils are one such example in which plants stay healthy in the presence of a high density of disease-causing organisms, although less is known about this phenomenon (Maldonado-Mendoza et al. 2009). Metagenomics provides a distinctive opportunity to investigate how microbial communities interact with crops to harness their potential for producing healthier and robust crops (Melcher et al. 2014). Metagenomic studies navigate the functional characteristics of belowground microbiomes and provide in-depth understanding over the genetically conserved rRNA gene-based phylogenetics (Cong et al. 2015; Rastogi and Sani 2011).

Complex soil microbial communities under diverse soil type and climate

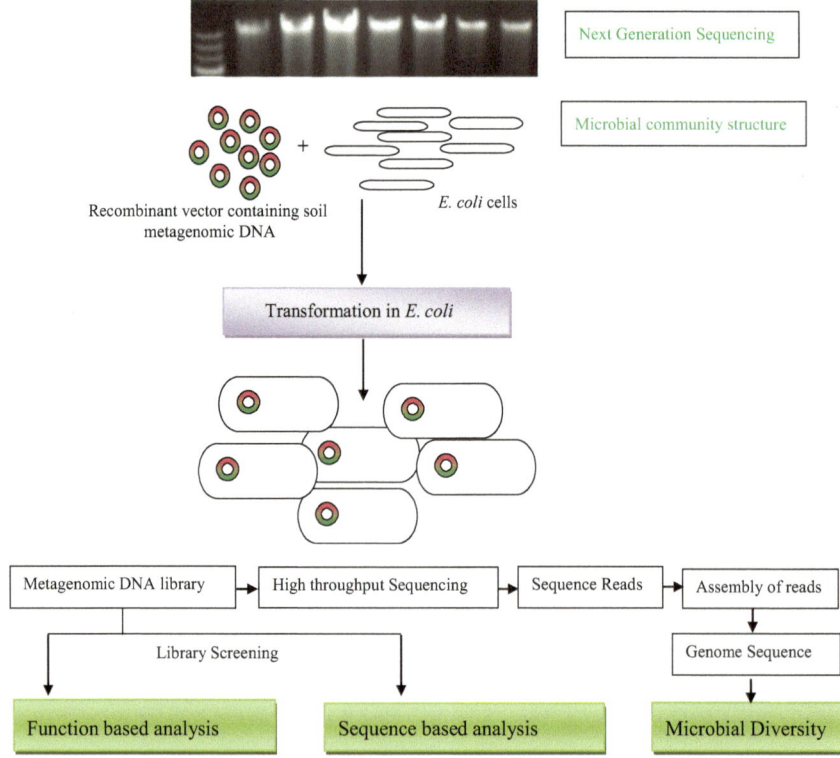

Fig. 4.2 Key processes for metagenomics DNA library preparation to explore the soil microbial community. In the sequential steps for development of the metagenomic DNA library from soil, isolated soil DNA is fragmented and subsequently cloned in to the desired vector such as a plasmid, cosmid, fosmid, or BAC (bacterial artificial chromosome) followed by transformation into the host cell. Further screening of desired clones with useful genes provides their ecophysiological role in the associated ecosystem. Microbial community structure can be predicted directly from soil metagenomic DNA via next-generation sequencing technologies

4.2.2 Metagenomics Data Analysis: Approaches and Challenges

The success of metagenomics is completely dependent on the high-throughput techniques for the processing of DNA from different environments and their sequence analysis after running on high-end sequencers. Furthermore, analyzing millions or trillions of the reads and their assembly to achieve a complete genome is a really challenging task (Aguiar-Pulido et al. 2016). Metagenomic projects face several challenges during the gathered data analysis, such as assembly, phylogenetic analysis, taxonomic binning, gene calling, and community-level analysis. Metagenomics invites many interesting computational problems such as gene prediction, sequence classification, and clustering, genome assembly, statistical comparison, functional annotation, microbial interactions modelling, etc. Data coverage is again a major challenge in the metagenomic study because of the poor availability of computational methods to estimate the magnitude of coverage (Rodriguezr and Konstantinidis 2014). Some of the major computational challenges include the assembly of the whole data, phylogenetic surveys, gene finding, and comparative metagenomic analysis for the metabolic pathways (Wooley and Ye 2009; Sharpton 2014; Mendoza et al. 2015; Filippo et al. 2012). In general, there are two approaches of computational metagenomics data processing and interpretation: (i) statistical analysis of functional metagenomic data and (ii) categorization of genes and from millions of the metagenomic reads (Sharon 2010). Metagenomic sequence data can inform us not only about the structural microbial communities in the soil or other habitats but can also assign functions to the microbes inhabiting the different habitats (Thomas et al. 2012; Sharpton 2014; Zhou et al. 2015).

4.2.2.1 Assembly

Assembly is mostly relevant to Sanger sequencing data wherein read length usually approaches 1000 bps. It is usually carried out using a single genome assembler. The use of comparative analysis with respect to template genomes may increase confidence in the final result. The transformation from the assembly of a single genome to the assembly of many genomes at once is not trivial and raises several issues. The presence of conserved regions in several different organisms is likely to harm the assembly because of the assembler's inability to differentiate closely related stretches. Such stretches are usually considered as being repetitive and are ignored. The same problem is caused by the presence of several copies of a region that belongs to an abundant genome, in particular when the region is highly polymorphic. This problem may be treated manually (Kunin et al. 2008; Miller et al. 2010; Teeling and Glöckner 2012). There exist certain tools for estimating the amount of gaps in single genome assembly for a certain amount of coverage. Constructing such models for multiple genomes is much more difficult and requires estimation of the diversity of organisms in the tested environment (Hooper et al. 2010; Ekblom and Wolf 2014; Sims et al. 2014). Assembly of next-generation sequencing (NGS) data is rarely done because of the short read length. Current assemblers for NGS data such as Velvet may handle de novo assembly of microbial genomes but are not suitable for eukaryotic genomes as well as metagenomes. Strategies and assemblers for NGS metagenomic data are, therefore, of great interest for the metagenomics community (Miller et al. 2010; Henson et al. 2012).

4.2.2.2 Phylogenetic Analysis

Phylogenetic analysis from the metagenomic data is again a challenge because the data are composed of short sequences from many organisms, raising several issues related to phylogenetic analysis, with respect to both traditional and metagenomic-specific analysis (Sharpton 2014; Kembel et al. 2011; Zhou et al. 2015). Locating an organism whose genome is known on the phylogenetic tree with respect to other known organisms is traditionally carried out using phylogenetic informative markers, primarily rRNA-encoding genes. More than 1,400,000 microbial 16S small subunit (SSU) rRNA coding genes can be found in such global databases as the Ribosomal-Database-Project for phylogenetic classification of environmental data (Rajendhran and Gunasekaran 2011; Cole et al. 2009; Sharon 2010). Other phylogenetic marker genes such as the recA and HSP70 are also used (Wu et al. 2014; Wu and Scott 2012). However, most reads and scaffolds in metagenomic projects obviously do not contain phylogenetic marker genes, which raises the need for other methods that may be based on oligonucleotide frequencies or other properties of the genome. Methods that make no use of universal genes may be useful also for phylogenetic analysis of viruses, in which no universal genes are available (Pride and Schoenfeld 2008; Iwasaki et al. 2013; Sharpton 2014; Darling et al. 2014). Even when reads with phylogenetic marker genes are available, the task of reconstructing a phylogenetic tree may not be trivial because only gene fragments are available. As a first step in the process, a multiple sequence alignment (MSA) of all sequences may be determined. Next, the MSA may be used for constructing a distance matrix for obtaining a phylogenetic tree (White et al. 2010). The fragmented nature of metagenomic data usually results in partial sequences of the same region, which makes it impossible to use current algorithms for MSA on such data. This problem may require a new type of algorithm for aligning and scoring multiple sequences (Sharpton 2014).

4.2.2.3 Taxonomic Binning

Taxonomic binning is another problem in metagenomics analysis. Sequence binning is a process for the separation of genome sequences into taxon-specific groups. A binning step may be part of the assembly process of metagenomic data or may be used for separating the genomes of a few members to study the biological processes. The two main approaches with respect to this problem employ comparative and nucleotide frequencies techniques, the latter being more successful (Thomas et al. 2012; Dröge and McHardy 2012; Sedlar et al. 2017). The TETRA sequence analysis software and its underlying similarity measure between sequences use a representation of tetranucleotide frequencies for a sequence, and the Pearson correlation coefficient as a similarity measure between sequences (Teeling et al. 2004). The PhyloPythia classification algorithm employs a support vector machine (SVM) for classifying sequences to known taxa, based on their nucleotide frequencies (Higashi et al. 2012).

4.2.2.4 Gene Calling

Gene calling is another challenge to be deciphered. Several good solutions have been proposed in reference to the issue of gene finding in fully sequenced genomes. Algorithms based on the Hidden Markov-Model (HMM) and Interpolated-Markov-Model (IMM) use sequence statistics for distinguishing between exonic and intronic regions. These algorithms are trained on sequences from the target genome; an initial set of genes is obtained by comparative methods. Approaches including search for translating gene segments (ORFs) by considering start and stop codons or aligning the genome against databases of known genes or proteins are well recognized (Salzberg et al. 1999; Mathé et al. 2002; Zhang 2002; Wang et al. 2004). All aforementioned approaches may encounter difficulties that result from the fragmented nature of the sequences in the projects. A large portion, sometimes up to 50%, of all the reads in metagenomic projects cannot be assembled with the rest of the sequences being assembled into short contigs of a few Kbps (Charuvaka and Rangwala 2011; Lin and Liao 2016). Considering the length of reads (400–1000 bps for Sanger and 454 sequencing, less than 100 for Solexa) and the fact that unicellular organism genes are usually small, devoid of introns and located approximately every 1000 bps, probably valuable information is present on most reads and contigs (Sharon 2010). Sequence statistics-based algorithms may encounter difficulties in analyzing such fragmented data as they depend on whole-genome statistics. Comparative-based methods are also expected to encounter difficulties from the many gene fractions (Ekblom and Wolf 2014). Certain approaches involve a clustering algorithm in which similar sequences in Global Ocean Sample (GOS) are clustered together on the basis of sequence similarity. The primary input for the algorithm is the pairwise sequence similarity between all sequences in the database computed using BLAST search. These similarity scores are used both for removing redundancies and also for the construction of core sets, which contain highly conserved sequences. Next, close core sets are unified based on profile–profile comparison. Last, the profiles of the resulting sets are used for sequence recruitment using PSI-BLAST. Clusters containing sequences that are similar to annotated sequences may be assigned predicted functionality; however, approximately 25% of the clusters in GOS do not have any known homologue (Rusch et al. 2007; Yooseph et al. 2007; Li et al. 2012).

4.2.2.5   Community-Level Analysis

Community-level analysis is another structural and functional issue of interest in answering questions related to microbial communities as a whole, rather than specific species of functional systems (Zarraonaindia et al. 2013). Community makeup is usually studied by analyzing phylogenetic marker genes, most notably the 16S SSU and 23S LSU rRNA coding genes. Species are identified on the basis of sequence alignment against databases such as RDP, and species richness is estimated based on statistical tools such as rarefaction curves (Sharon 2010; Guo et al. 2016). The functional attributes of the microbes are estimated based on the gene contents of the metagenome and are built by estimating the relative abundance of each gene or pathway (Kunin et al. 2008; Carr and Borenstein 2014). Identification of genes and pathways is usually done by aligning all sequences in the metagenome to databases such as the NCBI Clusters of Orthologs (COG) or the Kyoto-Encyclopedia developed for the Genes and Genomes (Randle-Boggis et al. 2016). Once the number of reads carrying each gene or pathway is determined, it is possible to use these numbers for estimating the comparative quantitative analysis of each function (Filippo et al. 2012).

4.2.3   Metagenomics Data Analysis: Available Tools and Techniques

A number of databases have enormous capability to store and analyze the metagenomic data. MG-RAST, IMG/M, and CAMERA represent three well-known metagenomics servers. MG-RAST (Keegan et al. 2016) provides a data repository along with a full-fledged platform for analysis with multiple comparisons. The EBI Metagenomics service has facilitated the users with an automated pipeline for the analysis and storage of metagenomic data and allows detailed analysis of the taxonomic relationships along with the functional and metabolic potential of any sample (Hunter et al. 2014b). Genomes-OnLine-Database (GOLD) renders knowledge concerning finalized and in progress microbial mega-projects across the globe (Mukherjee et al. 2017). Both IMG/M (Chen et al. 2017) and MG-RAST are well-organized platforms that allow users to insert their metagenomic data, compare the data with available databases, metagenomes, without requiring the end-user-based raw data. CAMERA (Seshadri et al. 2007) presents more flexible annotation schema through the user's needs to understand the data annotation and analytical pipelines sufficient for their interpretation. MEGAN is one more tool applied for analyzing annotated data obtained from BLAST analysis under functional or phylogenic study (Huson et al. 2007).

Many reference databases such as KEGG (Kanehisa and Goto 2000), eggNOG (Muller et al. 2010), COG (Tatusov et al. 2003), PFAM (Finn et al. 2014), and TIGRFAM (Haft et al. 2003) provide overall functional physiology of the metagenome. However, none of the reference databases covers entire functions of a particular system with significant and clear-cut interpretations. Taxonomic analysis under metagenomics by classifying a set of anonymous DNA reads is another area to resolve while investigating any soil entity (Fig. 4.3). Several computational tools were developed for this purpose: PhymmBL uses a hybrid approach combining an interpolated Markov model framework with similarity-based BLAST search (Brady and Salzberg 2009), MEGAN uses the lowest common ancestor of the top three nearest neighbours based on similarity search (Huson et al. 2007), and AMPHORA infers a phylogenetic tree based on a pair-wise similarity matrix. These tools indeed represent significant improvements for classifying DNA reads (Wu and Eisen 2008). For example, most of these tools are not good at handling novel taxonomic groups, require huge amount of computational power, and are not accurate enough.

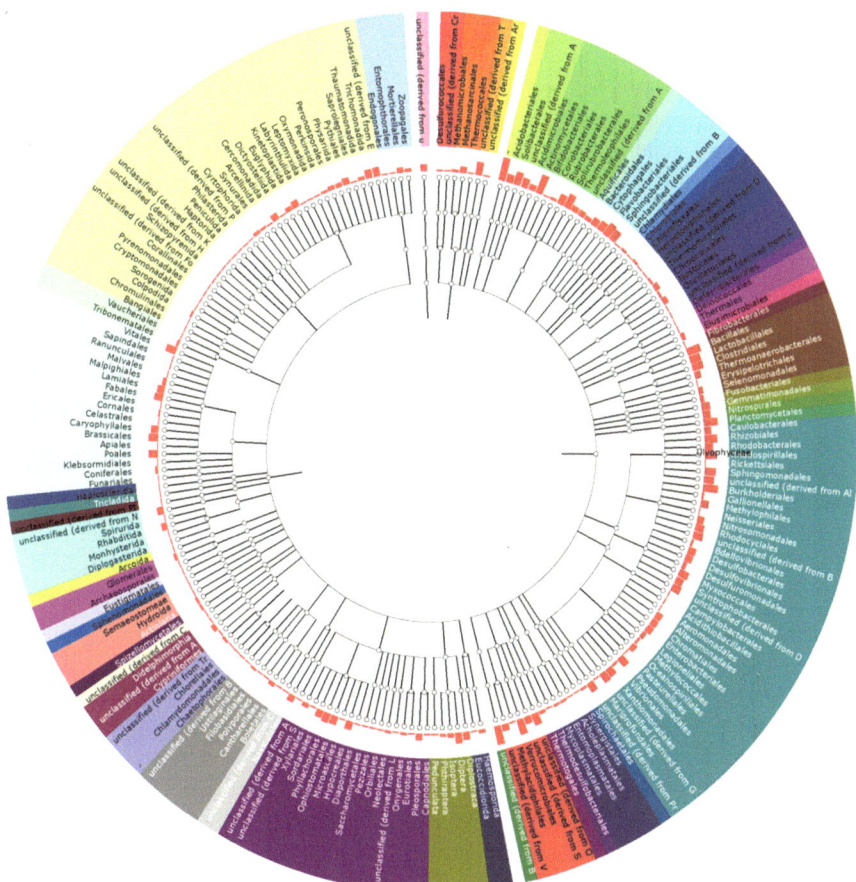

Fig. 4.3 Representation of metagenomic data after detailed bioinformatics analysis: taxonomic classifications using lowest common ancestor of a particular metagenomic dataset. Different *colours* show the particular structures of microbial diversity under soil system belonging to a specific phylogeny

Taxonomic binning is a core problem in metagenomics in which DNA reads are classified into taxonomic groups termed bins. To date, several binning algorithms are capable of classifying sequences into pre-defined bins based on either sequence or DNA signature similarity. Sequence similarity-based algorithms attempt to align the metagenomic data to a database of fully or partially sequenced genomes (e.g., the MEGAN software). The approach may provide accurate results when the binned sequence has close relative(s) in the genome database, but its accuracy and sensitivity drop significantly drop when this is not the case. DNA signature-based approaches use oligonucleotide frequencies to generate profile vectors for all sequences in the metagenome. Once generated, these vectors are compared to a set of predefined profile vectors representing known genomes and assigned to the closest genome (e.g., the PhyloPythia algorithm) (McHardy et al. 2007). These algorithms tend to be more accurate but may still fail for short read lengths. Most existing algorithms are, in fact, classification algorithms that require a "training set" consisting of sequenced genomes. These algorithms are likely to fail for sequences with no close relatives in the training set.

With the advent of metagenomics and computational biology, scientists all over the world have generated huge amounts of genomic data in the form of reads and contigs. Under the situation a variety of the bioinformatics tools has been developed with time to process, analyze, interpret, and annotate these huge amounts of metagenomics data. Furthermore, the precision and accuracy of the bioinformatic tools also created a big question mark; however, it has been managed with newly developed technologies. Under the topic of annotation, RAST (rapid annotation by subsystem technology) is a most commonly used server (more than 20,000 global users) for metagenomic data annotation and annotated approximately 70,000–80,000 (Overbeek et al. 2013). More detail about the server is well explained by Aziz et al. (2008). In the process of annotation, generally feature- and function-based analysis of coding regions has been carried out with the help of the various bioinformatics tools such as the gene finder, FragGeneScan (Thomas et al. 2012).

In Fig. 4.4, functional abundance in a particular metagenomic dataset is illustrated. Pathway reconstruction is a related problem in which common cellular or physiological processes in a genome or a metagenome are determined without estimating their abundance (Filippo et al. 2012). Figure 4.5 provides an example of pathway mapping through the KeggMapper tool of the MG-RAST server. Functional analysis at the pathway level is mainly used for two purposes: computation of pathway relative abundance, and pathway content comparison. Computing relative abundance of pathways within a single sample provides an overall view of the environment and was used in many studies and platforms. Comparing pathways' abundance between samples makes it possible to identify pathways that are enriched within one of the environments with respect to the other. Derivatives of pathway content comparison may be used for clustering functionally similar environments using metrics over pathway abundance vectors (Filippo et al. 2012). Some limitations, such as less-developed bioinformatics tools and assembly of a thousand sequence reads of the metagenomics analysis, still exist so that complex microbial communities may not appear in any assembly. High-throughput sequencing tech-

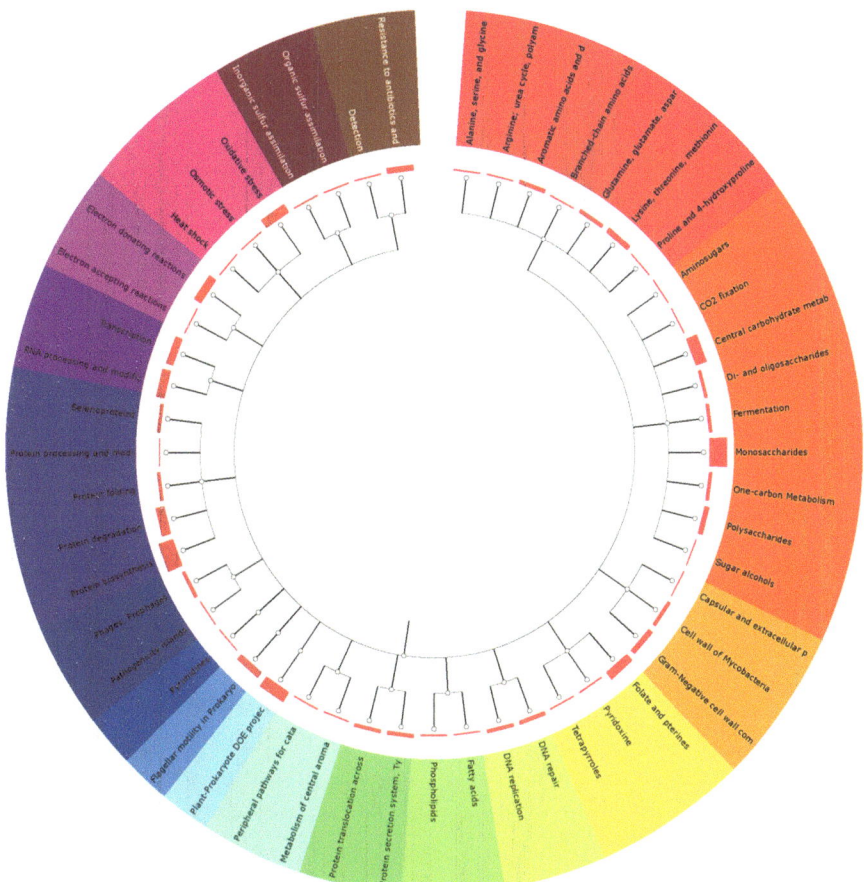

Fig. 4.4 Hierarchical classification of functional abundance in a particular metagenomic dataset through MG-RAST server: functional classifications of a particular metagenomic dataset after detailed bioinformatics analysis. Different colours show the particular functional group of microbial diversity under soil system belonging to specific soil processes and pathways

nologies have improved the capabilities of metagenomic studies to a greater strength but at the same time have led to the generation of large data sets that require high-end algorithms and computational tools for data analysis and storage. Analysis of the resulting data requires data mining approaches to find novel genes and gene families that can be connected with the functions of the microbial communities within the habitat. Metagenomics applicable to the group of integrated genomic cum in silico computational approaches that directly analyzes complete genome of belowground microorganisms and attempts to link them to corresponding functions (Rastogi and Sani 2011; Creer et al. 2016). This highly emerging field is now accountable for significant improvements in microbial studies related to their within and between interactions and evolvements under myriads of the environments. In the past decades, different research laboratories worldwide are now enthusiastically

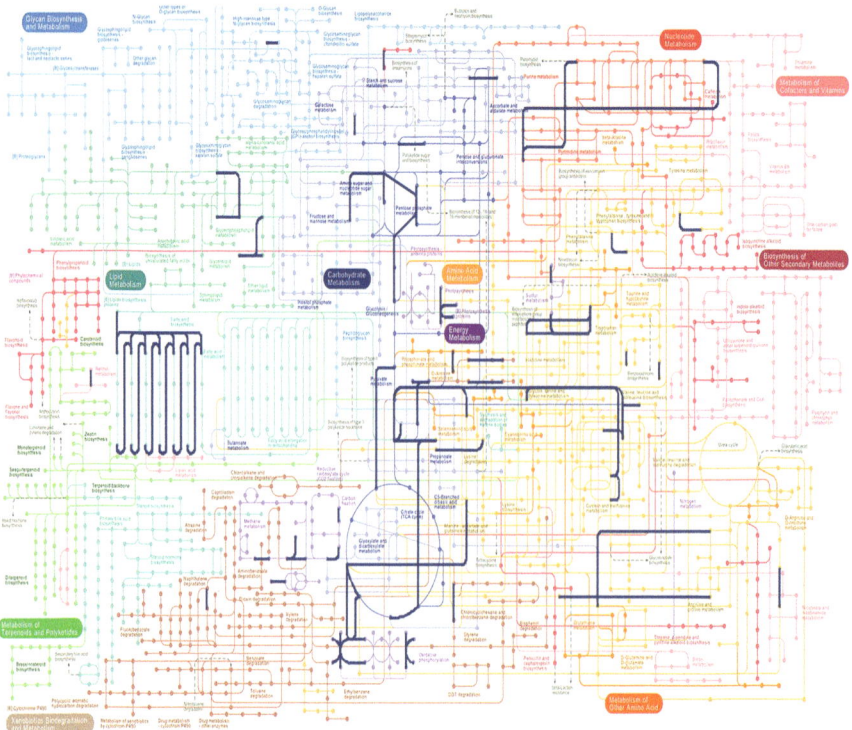

Fig. 4.5 Projection of a particular metagenomic dataset onto KEGG pathways through KeggMapper tool of MG-RAST server (dark blue lines represent query metagenomic dataset mapped over different pathways)

involved in these studies. With growing interest in deciphering the fate and behaviour of microorganisms in the environment and the increased number of activities on a plethora of methodological expertise, it has now become most demanding and challenging to extract accurate and reproducible knowledge from the metagenomic sequence information and link this knowledge to functions.

4.2.4 Metagenomics: Success Stories

Among the enormous applications of metagenomics, the most important ones include the exploration of environmental samples, the gut microbiome, unidentified microbes, genetic elements, underlying interlinked processes, cleanup of sewage and solid waste, and the implications of the huge amounts of meta-sequence data coming from the unseen microbial communities (Handelsman 2004; Bashir et al. 2014; Sebastian et al. 2013; Wang et al. 2016). Metagenomics in combination with the metatranscriptome and proteome can provide better insight into the soil microbial profile and also decipher many unidentified soil processes and mechanisms (Segata et al. 2013). Insights achieved from these metagenomic data of interlinking soil, air, and water systems are navigating the hidden mystery of associated systems and opening a new window for research for better human well-being by improving the process of disease identification, recognition of pathogenic contamination, industrial biotechnology (bio-prospecting, bio-remediation studies, etc.) (Handelsman 2004; Bashir et al. 2014). With the advances in precise metagenomic DNA isolation methodology, sequencing platforms, bioinformatics tools, and high-end computation biology, metagenomics is on track to solve the many global problems related to ocean sciences (oceanic microbiome), soil ecology (Earth microbiome goal), and human health (gut and human microbiome goal) seed microbiome (Turnbaugh et al. 2007; Tseng and Tang 2014; Gilbert et al. 2014; Berg and Raaijmakers 2018). All the aforesaid microbiome-level knowledge and development of pan-genome and the CRISPER-CAS-mediated modification system to alter the microbial functions have given new focus to the microbial genomics and surely will add multipurpose benefits to human health and sustainable global development (Zhao et al. 2018; Akinsemolu 2018; Shapiro 2017).

Chapter 5
Metatranscriptomics and Metaproteomics for Microbial Communities Profiling

Abstract Metatranscriptomics and metaproteomics are major breakthroughs of the next-generation sequencing technologies. Metatranscriptomics and metaproteomics not only provide information about the taxonomic structure of the microorganisms in soil but also provide information about their functional attributes and diversity. Gene expression under varying environmental conditions can be analysed by polymerase chain reactions and microarray. Similarly, techniques such as metatranscriptomics can be used for genome-wide gene expression analysis, providing novel insights about the ecology of the microorganism-mediated processes. In the present chapter we have highlighted the importance, benefits, challenges, process, and procedures of metatranscriptomics and metaproteomics for analysing microbial communities from diverse environments. Metatranscriptomics and metaproteomics have carried out significant revolutions in the field of microbial ecology via exploring the plant–microbe and microbe–microbe interactions.

Keywords Functional genomics · Microbial diversity · Metatranscriptome · Metaproteome · Microbial interactions

5.1 Metatranscriptomics

Metatranscriptomics concentrates on expressed genes in the entire microbial community and provides a view of the active functions of the community of microorganisms in a particular environment (Moran 2009). Metatranscriptomics is a very powerful technique for functional profile analysis and the study of complex microbial physiology and the structure of unknown microbial communities. The metatranscriptome represents the real-time expression of mRNA in selected samples by analysing the total captured mRNA (Poretsky et al. 2005). Expression analysis of the (meta)genome offers complete functional characterization of the microbiome (Carvalhais et al. 2012; Aguiar-Pulido et al. 2016). Recent developments in the 'omic' technologies have revolutionized the metagenomics field (Bashiardes et al. 2016b).

Over the years, development of massively parallel sequencing platforms has led to efficient transcriptome analysis that offers a wealth of information regarding gene expressions of microbial community systems from diverse environments (Wang et al. 2009). Thus, metatranscriptome analyses support metagenomics data by revealing accurately the genes which are transcribed (Franzosa et al. 2014) and allow analyzing functions from a potential collection of microbes (Bashiardes et al. 2016a, b). From these functional data, active metabolic pathways can be recognized and linked with the microbial communities in any specific environmental conditions. Thus, metatranscriptomics reflects a more informative view, disclosing particulars about the population which are transcriptionally active rather than identifying only genetic composition. This aspect is of extreme significance; it is elusive and clinically important to show dissimilarities in different groups of the active bacteria that frequently occur among human individuals (Franzosa et al. 2014; Bashiardes et al. 2016a, b). In this approach, sequencing of the mRNA can be done without using any primer or probe. Total RNA isolated from complex microbial samples from the soil is reverse transcribed into complementary DNA without polymerase chain reaction (PCR) application or molecular cloning. Consequently, pyrosequencing is done to produce various cDNA and rRNA tags. Taxonomic profiling is performed using MEGAN software and a specific rRNA reference database having sequences of small and large subunits of rRNA. Later mRNA tags produced by sequencing results revealed sequence-dependent transcriptomes of associate microbial groups. Community profiling can be performed using ribo-tags and consensus rRNA sequences only for certain taxa (Fig. 5.1).

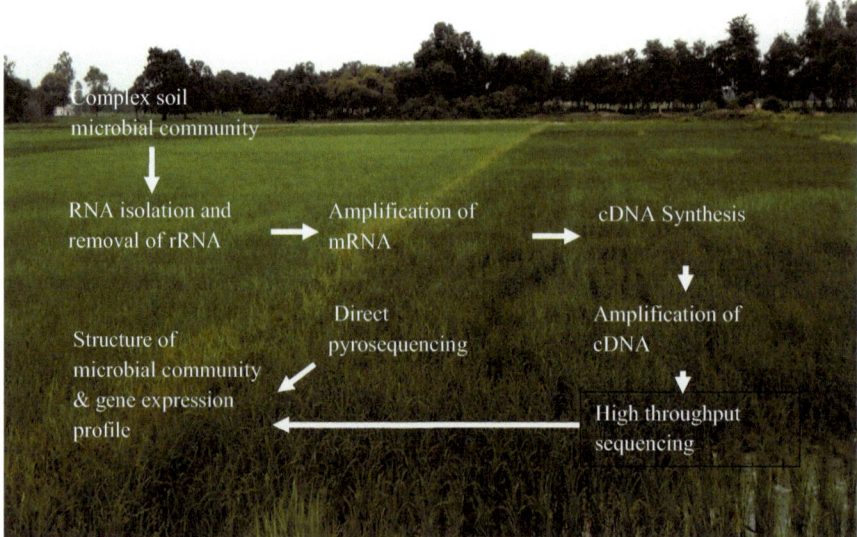

Fig. 5.1 General workflow for the soil metatranscriptomic analysis. Soil metatranscriptomics starts with total RNA isolation from the soil sample and the removal of rRNA to increase the purity of mRNA. This mRNA is reverse transcribed into cDNA and amplified to clone and generate the cDNA library. Finally, using pyrosequencing and reference database, structure and function of unknown microbial communities can be identified. Metatranscriptomics provides detailed insight about structure and functionality of the unculturable microbial world of the soil

5.1.1 Isolation and Processing of Microbiome mRNA

Metatranscriptomic analysis involves isolation of total RNA from microbial communities of a particular environmental sample. In eukaryotic samples, messenger RNA can be selected by cDNA mediated synthesis through application of oligo-d (T) primers that utilise the characteristic poly-A tail of mRNAs. Prokaryotes have only 1–5% of mRNA of the total RNA (Peano et al. 2013), even without the poly-A tail, rendering it unusable for the previously mentioned eukaryotic fashioned cDNA synthesis. Many technological advancements have been put into practice to offer a solution to this problem (Sultan et al. 2014; Sharma et al. 2010). Probe application specific to selective rRNA adhered to magnetic beads provides efficient removal of rRNA. The procedure engages probe annealing to selected sequences (rRNA) and subsequent magnetic removal of the rRNA (Sultan et al. 2014). Only an enriched population of other mRNAs that corresponds to transcriptionally active genes is used in the method. Furthermore, for massive parallel sequence generation, RNAs are fractionated for corresponding cDNA synthesis, and then cDNA ends are linked with adapters followed by end repair, finally generating a library for further amplification and sequencing purposes. Sequence reads are then processed for mapping and the expressed genes are recognized on the basis of the sequence reads covering these regions (Bashiardes et al. 2016a, b).

5.1.2 Computational Analysis of Metatranscriptomics Data

The sequenced metatranscriptome dataset possess millions of mRNA molecules, called RNA-seq reads. With increasing number and sample size of metatranscriptomic studies, highly efficient analysis platforms are required to draw meaningful inferences from these datasets (Gosalbes et al. 2011; Korf and Rehm 2013). Various extensive analysis suites such as HUMAnN (Abubucker et al. 2012) and MG-RAST (Glass et al. 2010) were developed and widely applied for efficiently resolving the problems related to interpretation of the raw data. These techniques are used in combination of specialized bioinformatics tools such as BOWTIE (Langmead and Salzberg 2012) and GEM (Marco-Sola et al. 2012) for mapping, Trimmomatic (Bolger et al. 2014) for quality filtering, and CuffDuff (Ghosh and Chan 2016) for differential gene expression, so that variations in the gene expression levels can be inferred from the raw sequenced mRNA reads (Bashiardes et al. 2016a, b). Certain analytical steps are necessary for this procedure and, as a result, they are consistently present in almost every metatranscriptomic. These steps are filtering of non-mRNA reads along with the host reads, low-quality reads trimming, open-reading frames (ORFs), mapping of the reads to a particular database, data normalization, and computing the expression levels of mRNA with alternative summarized statistics (Wang et al. 2009; Bashiardes et al. 2016a, b).

Assembling the reads into contigs is nonobligatory and they can be mapped to the reference genomes. The assembly process is computationally difficult and needs a high standard sequenced dataset; thus, it has the prospect of discovering new facts about mRNA expression that were not feasible earlier (Bashiardes et al. 2016a, b). In experimental terms, deep sequencing is needed to perform the assembly, and thus only the most abundant sequences can be organized from a larger set of reads (Morgan and Huttenhower 2014). A step of assembly is needed in the case of limitation of reference genome and gene annotation platforms. In cases when the reference genome is unavailable, annotations for the sequenced transcripts are generally acquired through sequence similarity searches to sequenced and annotated proteins. In other terms, alignment is carried out between the assembled transcripts and large annotated protein databases through software such as Blast2GO (Conesa et al. 2005), and when almost identical proteins are reported, then the parallel biological function is typically concluded. Trinity-like graph-theoretic concept-based advanced computational methods have been developed for reconstruction of a complete transcriptome (Grabherr et al. 2011).

Another critical problem in biological interpretation from metatranscriptomics data is to link the sequenced RNA dataset with corresponding DNA sequences. Simultaneous analysis of such datasets allows researchers to identify expressed mRNAs from the total present mRNAs. With the presence of the step of assembly for analysing the sequenced RNA and subsequent post-normalization, the data can be transformed to the corresponding expressed gene value, which can be further interpreted through statistically analysing the 16S-rRNA and metagenomic sequencing. This step can potentially reveal the expression level of mRNA, species richness, and similarity percentage of the samples (Bashiardes et al. 2016a, b).

5.1.3 Metatranscriptomics and Soil Microbial Diversity

Soil microbial communities carry out crucial ecosystem functions such as decomposition and geochemical cycling that robustly affect physical characteristics of the soils along with plant health and nutrition. Soils are complex and offer an enormous diversity of habitats owing to their structural features such as size, shape, and pore networks connectivity, with additional features such as the complication of resources, physicochemical characteristics, and biological interactions (Carvalhais et al. 2012). Metatranscriptomics allows in-depth information about the potential expression of genes at the sampling time. As post-transcriptional and post-translational gene expression performs protein synthesis regulation, control of gene expression at the transcriptional level allows microbes to quickly adapt to varying environmental conditions (Moran 2009). Thus, direct regulatory reactions to environmental changes may be better revealed by metatranscriptomics as compared to metaproteomics (Moran 2009; Carvalhais et al. 2012). Metatranscriptomics is used to study the diversity of soil microbes from different environmental samples such as arctic peat soils (Tveit et al. 2014), sludge (Yu and Zhang 2012), rhizospheres

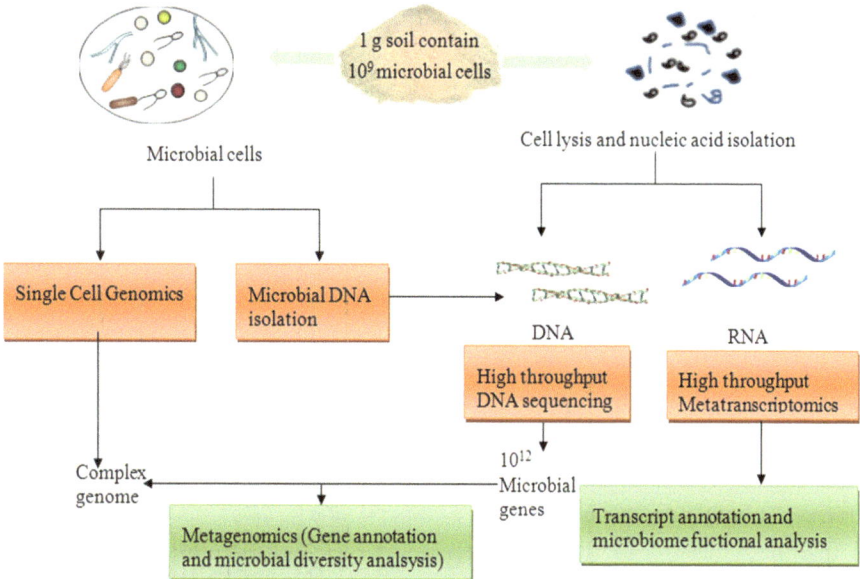

Fig. 5.2 Overview of metagenomics and metatranscriptomics analysis of the soil showing key steps in generating metagenomic and metatranscriptomic libraries per gram of soil sample. Richness of the microbial cells and genes varies with soil type; however, typical values are indicated here. Estimation of the total number of transcripts g⁻¹ soil is a very tedious task

(Chauhan et al. 2014; Molina et al. 2012), acid mine drainage (Chen et al. 2015), and forest soils (Damon et al. 2012) (Fig. 5.2).

5.2 Metaproteomics

An in-depth understanding of microbial ecology facilitates a clear picture of different processes such as biodegradation, decomposition, nutrient management, and global mineral recycling. Owing to this complexity, study of these communities imposes noteworthy challenges (Lacerda et al. 2007). Multiple approaches have been used for the analyses of microbial communities including traditional enrichment techniques, multispecies modelling, and lipid-, DNA-, and RNA-based approaches (Dumont and Murrell 2005; Ogunseitan 2005; Sharkey et al. 2004; Lacerda et al. 2007). Despite these technologies, information about different functions such as metabolic capacity, population dynamics, and physiological responses to varying environmental conditions is still unexplored (Lacerda et al. 2007). Proteomics possesses potentials to facilitate functional information in the microbial communities through its application to complicated communities of unknown, uncultured microbes (Lacerda et al. 2007). Proteomics facilitates the detection and quantification of expressed genes, providing understanding about how the metabolic machinery used by microbes under diverse habitats as proteins are considered as signature molecules of the cellular processes (Hettich et al. 2013).

Proteomics is stated as a wide-scale study of proteins synthesized by any living entity (Wilkins et al. 1995). It emerged through 2-D gel electrophoresis-mediated mapping of protein expression (O'Farrell 1975). The application of 2-D gel electrophoresis made isolation of protein and peptides feasible from a cellular mix of diverse complexity (Pedersen et al. 1978). Identification of protein is a tedious process because of the limited availability of responsive and high-throughput sequencing tools (Maron et al. 2007). After the 1990s, proteomics emerged as an advanced tool because of the availability of highly effective mass spectrometry (MS) for peptide ionization, permitting quick and extremely robust identification and characterization of proteins (Pandey and Mann 2000; Yates et al. 1993). Along with this, there was parallel progress in the field of bioinformatics tools for the analysis of 2D gels and mass spectra, allowing (1) quick detection of proteins through database matching with available mass spectra (Pandey and Lewitter 1999), (2) further reverse genetics-based identification of corresponding genes (Mann and Pandey 2001), and (3) the assessment of protein posttranslational modifications (Anderson et al. 2002). Apart from these technological advancements, most of the proteomics studies were conducted under laboratory conditions and do not account for the diverse interactions occurring among the microbial communities. Thus, modified approaches are required for the in situ identification and evaluation of the wide-scale protein expression at the level of population or community (Maron et al. 2007).

On the basis of the DNA extraction from a specific environment, development of microbial community genomics represents a path forward (Martin et al. 2006; Venter et al. 2004; Tyson et al. 2004; Lacerda et al. 2007), although apart from the discovery of new genes, not much information was related to function and system dynamics. In contrast, proteomics studies facilitate information about rapid physiological responses as proteins are synthesized and folded within seconds (Lehninger 1965; Lacerda et al. 2007). Limitations of the DNA- or RNA-based approaches were recognized at the later stages of the postgenomic era by the production of structural rather than functional information from these approaches (Maron et al. 2007). Large-scale proteomics studies of indigenous microbial communities, that is, metaproteomics, have emerged as a valuable approach to know the functionality of the microorganisms in the ecosystem. It is expected that thorough metaproteome characterization can help in exploring the structural-functional diversity of microbes from different environmental components (Maron et al. 2007).

5.2.1 Metaproteomics: Benefits and Challenges

Metaproteomics is the characterization of the whole protein component of environmental microbial samples (Wilmes and Bond 2004; Tanca et al. 2014). Analysis of metaproteome datasets reveals the community structure-function and dynamics of the microorganisms in a particular environment. It further improves our understanding about microbial recruiting, organism interactions, cooperation and competence for nutrient resources, and metabolic activities (Hettich et al. 2013). At a somewhat higher level, the information is crucial for depiction of host–microbe crosstalk such as microorganism–plant or microorganism–human interactions (Hettich et al. 2013). In a nutshell, environmental metaproteomics perspectives have founded a "proof-of-concept" that can facilitate applications to numerous significant research areas such as bioremediation, carbon cycling, bioenergy, and human health (Hettich et al. 2013). Until now, most of the proteomics experiments are focused on the analysis of single species under different conditions, leaving the perspective of microbial community proteomics almost totally unexplored (Lacerda et al. 2007). Furthermore, metaproteome analysis from different environmental conditions allows (1) identification of novel genes and cellular pathways and (2) recognition of stress-responsive proteins (Maron et al. 2007). There exist numerous challenges in microbial community proteomics such as representative protein extraction, sample complexity for separation (including even thousands of bacterial proteomes), and an almost total absence of genomic sequences for microbes in ecological communities (Lacerda et al. 2007). Metaproteomics analysis of diverse environments such as the natural microbial biofilm (Ram et al. 2005), water (Kan et al. 2005; Ogunseitan 1993, 1996, 1997), sediment (Ogunseitan 1993), soil (Ogunseitan 1993; Singleton et al. 2003), and processed sludge (Wilmes and Bond 2004) have led to the discovery that the protein complement of the metagenome was much more complex and variable depending on the target environment (Ogunseitan 1993) and surroundings (Ogunseitan 1996; Singleton et al. 2003).

5.2.2 Metaproteomics Approaches

Metaproteome analysis involves various technical processes such as microbial protein extraction from the samples and subsequent identification and functionality assessment of the protein networks. Similar to in situ nucleic acid-based approaches, in metaproteomics analysis also the most critical process is making sure that the isolated proteins truly represent the sample (Maron et al. 2007). The success of any metaproteome experiments depends upon three factors: (i) efficient protein isolation from the sample, (ii) protein segregation and fractionation before detection, and (iii) robust protein identification (Hettich et al. 2013). It is largely applicable in the case of proteomic analysis of the environment as their highly heterogeneous and complex nature makes the specific isolation of the microbial proteome difficult. The extraction methodology differs according to selected proteins (which could potentially be from prokaryotes or eukaryotes, cellular or extracellular in nature) and by the successive techniques of proteome study (i.e., comparable 2-D protein maps, identification of proteins and enzyme activities) (Maron et al. 2007).

After the isolation of protein samples, diverse biochemical approaches are employed for the metaproteome study depending upon the required resolution and information. To obtain a "proteofingerprint" of the microorganisms, environmental proteins can be isolated by normal or two-dimensional (2-D) gel electrophoresis. 2-D gel electrophoresis offers better segregation of proteins, and these proteins can be further analysed by a mass spectrometric analysis-based database search; however, the approach has its own limitations such as the infeasibility of consistently monitoring low abundance (Gygi et al. 2000), and extremely hydrophobic, extremely acidic, or extremely basic proteins (Lee 2001). There is considerable development of numerous technologies for proteomics studies, such as chromatography or/capillary electrophoresis separation (Lee 2001; Yates 2004) and protein microarray developments (Ramachandran et al. 2004). Protein microarray allows robust studies of protein mixtures and promotes its application for analysis of samples from diverse environments. After identification of the proteome from particular environments, they are linked to their genes through reverse genetics, which is a key objective of metaproteomics. However, the technological advances are still inadequate and need optimization (Maron et al. 2007; Schulze et al. 2004). Many physicochemical methods have been reported for protein extraction, depending on proper cell lysis of the buffer application using detergents (SDS, CHAPS, Triton X-100) (Wilmes and Bond 2004; Chourey et al. 2010), chaotropic agents (urea, guanidine hydrochloride) (Wilmes and Bond 2004; Verberkmoes et al. 2009), reducing agents [dithiothreitol (DTT), tributylphosphine] (Kan et al. 2005), and other compounds (phenol, NaOH) (Keiblinger et al. 2012; Leary et al. 2012; Benndorf et al. 2007), along with heat treatment (Singleton et al. 2003; Ogunseitan 1997; Schneider et al. 2012), mechanical disruption (Wilmes and Bond 2004; Keiblinger et al. 2012; Kolmeder et al. 2012), and sonication (Tang et al. 2014). Once the protein is isolated, compounds that may hinder the digestion, chromatographic separation, or mass spectrometric studies require urgent removal. Generally, this is accomplished by precipitating the protein by addition of acetone, trichloroacetic acid, and other compounds to the cellular extract (Chourey et al. 2010; Benndorf et al. 2007; Leary et al. 2012), and the pelletized protein is further resuspended in buffer (Fic et al. 2010; Jiang et al. 2004). An additional efficient approach is to carry out one-dimensional electrophoresis separation of proteins and their subsequent digestion, which helps in capturing the hindrance causing chemicals in the gel and provides protein in sliced gels (Kolmeder et al. 2012; Ferrer et al. 2013). However, despite its efficiency, the approach is a time- and labour intensive process with lower reproducibility, thus less preferred (Choksawangkarn et al. 2012). One of the latest alternatives is the filter-aided sample preparation (FASP), which offers better cleanup and enzymatic cleavage, done in a ultracentrifugal filter (Wiśniewski et al. 2009). It was successfully applied in microbe samples and reported to be better than other approaches, accounting for less protein content in the samples (Tang et al. 2014). Moreover, sample complexity should be reduced for better insight through MS studies. In earlier metaproteomics studies, it was achieved by fractionation of proteins (Kolmeder et al. 2012; Perez-Cobas et al. 2013) or peptides (Verberkmoes et al. 2009; Schneider et al. 2012; Ram et al. 2005). However, additional fractionation requires more sample concentration, increased MS time length, with tougher reproducibility; for example, the preferred approach, that is, 2D-LC-MS, is technically challenging and demands longer time duration for analysis (Köcher et al. 2012; Verberkmoes et al. 2009).

Recently, introduction of another simple process based on NanoLC-ESI-MS/MS allows detection of more than thousands of proteins in a single reaction (Köcher et al. 2012; Yu et al. 2014; Tanca et al. 2014). It is imperative to have the complete genome sequence of the sample under consideration for proteogenomic comparison of obtained protein/peptide data for identification and characterization (Hettich et al. 2013). Proteogenomics act as a connecting link between genomics and proteomics (Beverley et al. 2002; Hettich et al. 2013). The technical necessities for proteomic studies are robust processing, efficient protein/peptides detection, wide range, capacity of complex sample analysis, and structural determination of the peptide sequences that are largely offered by mass spectrometry (MS) (Hettich et al. 2013). Initial studies in proteomics were conducted with 2-D gel electrophoresis (Klose 1975; O'Farrell 1975), frequently followed by MS detection, which is supported by use of multiple dyes in gel electrophoresis (Unlu et al. 1997). The capacity to interface multidimensional liquid chromatographic separation with MS allowed better insight in intricate samples (Delahunty and Yates 2007; Peng et al. 2003). The introduction of robust MS methods, such as the quadrupole time-of-flight mass spectrometry (Q-TOF-MS), linear trapping quadrupole-Fourier transform ion cyclotron resonance-MS (LTQ-FTICR-MS), and linear-trapping-quadrupole (LTQ)-Orbitrap-MS led to enhanced ability for quick scanning and better resolution of a wide range of protein/peptide mass, allowing high-throughput proteomic analysis (Hettich et al. 2013). Computational analysis, optimized sample processing accompanied by MS identification, seeks special attention in proteomics studies (Tanca et al. 2014). The efficacy of the method appropriate for the shotgun proteomic study of intricate samples depends basically on these processes:

- Protein isolation: an all-inclusive protein content of the complete set of microbes from the sample is required
- Cleanup: detergents should be used for removing hindrance-causing compounds before digestion of proteins
- Pre-fractionation: effective separation of peptides/proteins before MS identification to reduce sample complexity and enhance analysis depth (Tanca et al. 2014).

5.2.3 Metaproteomics and Soil Microbial Communities

Soil is highly heterogeneous in nature and harbours a highly diverse vast microbial biomass, more so than its other counterparts in the environment (Mocali and Benedetti 2010; Keiblinger et al. 2012).

- Metaproteomics study of the soil environment is an audacious work as it provides a wealth of information about soil proteins. Sample processing is the critical step in the proteome analysis for better resolution (Wang et al. 2006). However, there are many technical challenges in soil metaproteomics studies because the samples are intricate (Bastida et al. 2009; Nannipieri and Smalla

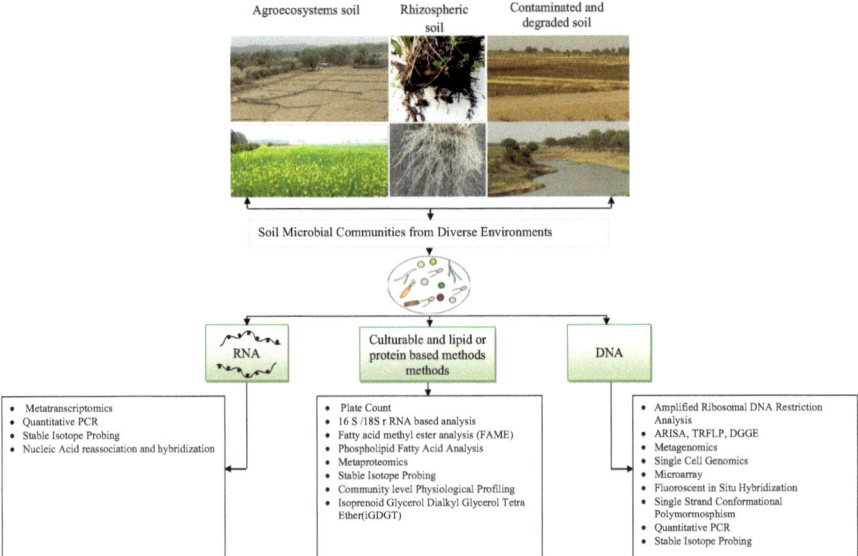

Fig. 5.3 General overview of the methods applied for studying the structure and function of soil microbial communities

2006), such as (i) lower soil protein levels, (ii) heterogeneity and spatiotemporal variations of soil microorganisms, and (iii) the presence of soil enzymes and humic colloids in soil (Nannipieri and Smalla 2006).

- Metaproteomics have promising applications in studying the microbial activity and interactions in soil (Keiblinger et al. 2012), although protein extraction from the soil is hard to achieve because of the inherent intricacy of soils. SDS-phenol- and SDS-NaOH-mediated approaches are most often used for protein isolation from soil (Keiblinger et al. 2012). Until now, we have had limited success in linking the soil metaproteome to microbial gene expression patterns (Bastida et al. 2009) along with the distinctive role of microbes in soil nutrient cycling (Hettich et al. 2010). To resolve these issues, we need to address soil heterogeneity and hydrophobicity (Hettich et al. 2010), lower protein levels, microbial richness, humic acid interference with protein isolation (Giagnoni et al. 2012), and thus the difficulties in proper isolation of the whole soil proteome (Williams and Taylor 2010). Overall, metaproteomics is a relatively young field of research and also there is huge diversity among the microbial communities in their biochemistry and structural-functional complexity, and thus considerable research efforts are needed for improvement, optimization, and standardization of sample preparation workflows for metaproteome studies. Overall highlights of all methodology for exploring microbial diversity are depicted in Fig. 5.3.

Chapter 6
Bioinformatics Tools for Soil Microbiome Analysis

Abstract Metagenomic approaches aid in exploring the structural and functional diversity of soil microorganisms. Sequence analysis of the large amount of data generated from soil microbial communities sequencing is a challenging issue. It is made feasible through bioinformatics tools, which provide sequence pipelines for the high-throughput screening of the soil metagenomic libraries. Such sequence analysis of metagenomic datasets reveals the genetic structure, gene prediction, proposed functions, and metabolic pathways of the analyzed microbial communities. Bioinformatic tools provide statistical procedures not only for comparison of metagenomic libraries but also to report the sampling and library creation artifacts. Here we discuss the bioinformatics tools for accessing the metagenomic information and platforms for data storage within databases (GenBank env), access, synthesis, and analysis.

Keywords Bioinformatics · Functional diversity · Gene prediction · Metagenomic · Structural diversity · Microbial taxonomy · Metabolic pathways

Metagenomics and metatranscriptomics data analysis and interpretation includes different steps: assembly and annotation, taxonomic and functional attributes, intra- and inter-environmental interaction networks, single-cell sequencing, simulation studies, and statistical analysis of downstream processes (Segata et al. 2013). Different tools and software employed at different steps of meta-omics data analysis and interpretations are provided in Table 6.1, and a brief explanation is also provided here.

6.1 Tools for Assembly and Annotation

(i) *Genovo:* Genovo is utilized for the de novo assembly of the genetic sequence that finds similar sequence chemistry under the platform. It tells us about the probable idea of read generation from the environmental sample. On

© The Author(s), under exclusive license to Springer Nature Switzerland AG 2020 61
R. K. Dubey et al., *Unravelling the Soil Microbiome*, SpringerBriefs in Environmental Science, https://doi.org/10.1007/978-3-030-15516-2_6

Table 6.1 Details of different tools and software employed in soil microbial community analysis

Sample no.	Name	Weblink	References
Tools for assembly and annotation			
1.	Genovo	–	Laserson et al. (2011)
2.	khmer	–	Pell et al. (2012)
3.	Meta-IDBA	http://www.cs.hku.hk/~alse/metaidba	Peng et al. (2011)
4.	MetAMOS	https://github.com/treangen/MetAMOS	Treangen et al. (2013)
5.	MetaVelvet	–	Namiki et al. (2012)
6.	MOCAT	http://www.bork.embl.de/mocat/	Kultima et al. (2012)
7.	SOAPdenovo	–	Li et al. (2010)
8.	MetaORFA	–	Ye and Tang (2009)
9.	FragGeneScan	–	Rho et al. (2010)
10.	MetaGeneAnnotator (MGA)	–	Noguchi et al. (2008)
11.	Orphelia	http://orphelia.gobics.de	Hoff et al. (2009)
12.	SILVA	–	Pruesse et al. (2007)
13.	Greengenes	http://greengenes.lbl.gov	DeSantis et al. (2006)
14.	Ribosomal Database Project (RDP)	http://rdp.cme.msu.edu/	Cole et al. (2009)
Tools for taxonomic profiling			
1.	AMPHORA2	http://wolbachia.biology.virginia.edu/WuLab/Software.html	Wu and Scott (2012)
2.	CARMA3	http://webcarma.cebitec.uni-bielefeld.de	Gerlach and Stoye (2011)
3.	ClaMS	–	Pati et al. (2011)
4.	DiScRIBinATE	–	Ghosh et al. (2010)
5.	INDUS	http://metagenomics.atc.tcs.com/INDUS/	Mohammed et al. (2011)
6.	MARTA	–	Horton et al. (2010)
7.	MetaCluster	–	Wang et al. (2012)
8.	MetaPhlAn	http://huttenhower.sph.harvard.edu/metaphlan/	Segata et al. (2012)

9.	MetaPhyler	http://metaphyler.cbcb.umd.edu	Liu et al. (2011)
10.	MTR	http://cs.ru.nl/gori/software/MTR.tar.gz	Gori et al. (2011)
11.	NBC	http://nbc.ece.drexel.edu	Rosen et al. (2011)
12.	PaPaRa	http://www.exelixislab.org/software.html	Berger and Stamatakis (2011)
13.	PhyloPythia	–	Patil et al. (2012)
14.	Phymm	–	Brady and Salzberg (2011)
15.	RAIphy	–	Nalbantoglu et al. (2011)
16.	SOrt-ITEMS	http://metagenomics.atc.tcs.com/binning/SOrt-ITEMS	Monzoorul Haque et al. (2009)
17.	SPHINX	http://metagenomics.atc.tcs.com/SPHINX/	Mohammed et al. (2011)
18.	TACOA	–	Diaz et al. (2009)
19.	Treephyler	http://www.gobics.de/fabian/treephyler.php	Schreiber et al. (2010)
Tools for functional profiling			
1.	HUMAnN	–	Abubucker et al. (2012)
2.	metaSHARK	http://bioinformatics.leeds.ac.uk/shark/	Hyland et al. (2006)
3.	Predicted Relative Metabolic Turnover (PRMT)	–	Larsen et al. (2011)
4.	RAMMCAP	http://tools.camera.calit2.net/camera/rammcap/	Li (2009)
Tools for interaction networks			
1.	SparCC	https://bitbucket.org/yonatanf/sparcc	Friedman and Alm (2012)
2.	CCREPE	–	Faust et al. (2012)
Tools for statistical tests			
1.	Metastats	–	Paulson et al. (2011)
2.	LefSe	http://huttenhower.sph.harvard.edu/lefse/	Segata et al. (2011)
3.	ShotgunFunctionalizeR	http://shotgun.zool.gu.se	Kristiansson et al. (2009)

(continued)

Table 6.1 (continued)

Sample no.	Name	Weblink	References
4.	SourceTracker	–	Knights et al. (2011)
Tools for single-cell sequencing analysis			
1.	IDBA-UD	http://www.cs.hku.hk/~alse/idba_ud	Peng et al. (2012)
2.	SmashCell	http://asiago.stanford.edu/SmashCell	Harrington et al. (2010)
Simulators tools			
1.	GemSIM	–	McElroy et al. (2012)
2.	MetaSim	–	Richter et al. (2008)
General toolkits			
1.	CAMERA	http://camera.calit2.net	Seshadri et al. (2007)
2.	IMG/M	http://img.jgi.doe.gov/m	Markowitz et al. (2012)
3.	MEGAN	–	Huson et al. (2007)
4.	METAREP	http://www.jcvi.org/metarep	Goll et al. (2010)
5.	MG-RAST	http://metagenomics.anl.gov	Meyer et al. (2008), Wilke et al. (2016)
6.	SmashCommunity	http://www.bork.embl.de/software/smash	Arumugam et al. (2010)
7.	STAMP	http://kiwi.cs.dal.ca/Software/STAMP	Parks et al. (2014)
8.	VAMPS	http://vamps.mbl.edu	Huse et al. (2014)
9.	EBI metagenomics	http://www.ebi.ac.uk/metagenomics/	Hunter et al. (2014b)

comparison with other short read assembly programs, Genovo access additional nucleotides bases and identifies/predicts more genes and provides a higher score for gene assembly (Laserson et al. 2011).

(ii) *khmer:* khmer is a memory-efficient graph representation for analysis of the *k*-mer analysis of metagenomically processed samples that overcomes the limitation of high-end-memory necessities for de novo assembly of short-reads sequenced via shotgun methods. Over a soil metagenome assembly, this method obtains highly efficiency from memory perspectives for assembly (Pell et al. 2012).

(iii) *Meta-IDBA* is a useful asset to assemble the diverse metagenomic reads possessing numerous genomes from various species. Meta-IDBA toolkit is available from http://www.cs.hku.hk/~alse/metaidba (Peng et al. 2011).

(iv) *metAMOS*: MetAMOS is a fully automatized and powerful tool to assemble the reads, produces a valuable scaffold, intron–exon identification, and complete annotation. MetAMOS is available from https://github.com/treangen/MetAMOS (Treangen et al. 2013).

(v) *MetaVelvet:* MetaVelvet provides an advanced version of Velvet with the power to assemble complex short reads of various sources. MetaVelvet is also able to reconstruct comparatively poor coverage reads (Namiki et al. 2012).

(vi) *MOCAT:* MOCAT is an extremely flexible and rapid platform for quality control; mapping and assembly of Illumina-based single or paired-end reads. MOCAT can be accessed on http://www.bork.embl.de/mocat/ (Kultima et al. 2012).

(vii) *SOAPdenovo*: SOAPdenovo is a cheaper approach for assembling of the large genomes via the de novo method (Li et al. 2010).

(viii) *MetaORFA:* Metagenomic ORFome Assembly (MetaORFA) facilitates metagenomic data analysis in three steps. In the first step, reads from the metagenomics dataset are annotated with putative protein-encoding regions. Then, the predicted regions are assembled through EULER assembly approach into a group of peptides. In the third, the processed proteins provide homologues and successive diversity with the help of the online databases (Ye and Tang 2009).

(ix) *FragGeneScan:* FragGeneScan is a Markov model-based improved gene prediction tool help in identification of the translated region of the genome from small reads (Rho et al. 2010).

(x) *MetaGeneAnnotator (MGA):* MetaGeneAnnotator (MGA) is again a potential tool for unicellular prokaryote gene identification (Noguchi et al. 2008).

(xi) *Orphelia:* Orphelia is also a gene prediction tool applicable in small reads that can be found on http://orphelia.gobics.de. (Hoff et al. 2009).

(xii) *SILVA:* Ribosomal RNA (rRNA) genes sequencing is at present a highly accepted method for assessing the nature of unculturable microbes. SILVA is a central all-inclusive web resource for most recent, quality controlled databases of aligned highly conserved sequences from all prokaryotes as well as complex microorganisms (Pruesse et al. 2007).

(xiii) *Greengenes:* Green genes are a 16S rRNA gene database frequently utilized for identifying chimeras, providing a platform for alignment and evolutionary linkage. It is online accessible from http://greengenes.lbl.gov as reported earlier (DeSantis et al. 2006).

(xiv) *RDP database:* The Ribosomal-Database-Project facilitates researchers for quality-control analysis of bacteria and archaea on the basis of highly conserved gene sequences. RDP is online available from http://rdp.cme.msu.edu/ and explained well (Cole et al. 2009).

6.2 Tools for Taxonomic Profiling

(i) *Amphora2:* AMPHORA2 is an automated phylogenetic inference approach that can be directly applied for high-throughput and high-quality genome tree building and metagenomic phylotyping. AMPHORA2 is available at http://wolbachia.biology.virginia.edu/WuLab/Software.html (Wu and Scott 2012)

(ii) *CARMA3:* CARMA3 is a tool for community structure identification from processed and unprocessed metagenomic reads and is compatible with many homologies searches. It is online available on following web-link http://webcarma.cebitec.uni-bielefeld.de (Gerlach and Stoye 2011).

(iii) *ClaMS:* Classifier for Metagenomic Sequences, abbreviated as ClaMS is a Java-enabled desktop application for binning of assembled contigs in metagenomic datasets through user-specified training sets and initial parameters (Pati et al. 2011).

(iv) *DiScRIBinATE:* DiScRIBinATE (Distance-Score Ratio for Improved Binning And Taxonomic Estimation) is a fast and accurate binning method for navigating the evolutionary relationship within the huge metagenome (Ghosh et al. 2010).

(v) *INDUS:* INDUS is a fast binning tool, capable of quickly categorizing massive reads under modelled and real data with better efficacy than composition approaches. The INDUS tool can be accessed at http://metagenomics.atc.tcs.com/INDUS/ (Mohammed et al. 2011).

(vi) *MARTA:* MARTA provides a Java platform for studying the evolutionary relationship among the gene sequences and depends on NCBI tools such as BLAST software and the Taxonomy database (Horton et al. 2010).

(vii) *MetaCluster:* MetaCluster is an unsupervised binning tool that can be utilized for the effective and precise grouping of small reads among myriads of species (Wang et al. 2012).

(viii) *MetaPhlAn:* MetaPhlAn is a metagenomic phylogenetic analysis tool that utilizes clade-specific marker genes for definite assignment of reads to microbial clades in a fast and accurate manner. It is online available from http://huttenhower.sph.harvard.edu/metaphlan/ (Segata et al. 2012).

(ix) *MetaPhyler:* MetaPhyler is tool for predicting the phylogenetic relationship and employs phylogenetic marker genes as a taxonomic reference. Accessible from http://metaphyler.cbcb.umd.edu (Liu et al. 2011).

(x) *MTR:* MTR performs clustering through multiple taxonomic ranks instead of lowest common ancestor (LCA). Rather than LCA, this precedes the reads one at a time; MTR take advantage of information shared by reads. It can be reached on http://cs.ru.nl/gori/software/MTR.tar.gz (Gori et al. 2011).

(xi) *NBC:* NBC is an online database employing the naïve Bayes classification system for classification of entire metagenomic reads to their best taxonomic match. Publicly available at http://nbc.ece.drexel.edu (Rosen et al. 2011).

(xii) *PaPaRa*: PaPaRa is a novel phylogeny-aware alignment procedure online available from http://www.exelixis-lab.org/software.html. It can be applied for alignment of small genomic reads with the reference phylogeny (Berger and Stamatakis 2011).

(xiii) *PhyloPythia:* This is a phylogenetic tool that provides precise and useful information about the generic clades of unidentified soil microorganisms (Patil et al. 2012).

(xiv) *Phymm:* Phymm is another improved taxonomic classifier that categorises very small metagenomic reads on the basis of the evolutionary background (Brady and Salzberg 2011).

(xv) *RAIphy:* RAIphy is a good binning and phylogenetic tool that is used commonly for processing and categorization of the metagenomic reads along with its efficient functional characterization (Albantoglu et al. 2011).

(xvi) *SOrt-ITEMS:* SOrt-ITEMS is a binning algorithm that employs alignment parameters rather than the bit-score to find the phylogenetic relationships. It can be downloaded from http://metagenomics.atc.tcs.com/binning/SOrt-ITEMS as mentioned by Monzoorul Haque et al. (2009)

(xvii) *SPHINX:* SPHINX is a mixed binning methodology that has good binning potential through employing both 'composition'- and 'alignment'-based binning algorithms. SPHINX is capable of analyzing metagenomic sequences as quickly as composition-based algorithms and as efficiently as alignment-based algorithms. It is online available from http://metagenomics.atc.tcs.com/SPHINX/ (Mohammed et al. 2011).

(xviii) *TACOA:* TACOA is a precise composite phylogenetic tool for ecological metagenomic datasets with the potential to identify the taxonomy of large as well as small reads (800 bp). It is transparent, rapid, and precise, and the reference set can be simply updated with the availability of new genome sequences (Diaz et al. 2009).

(xix) *Treephyler:* Treephyler allow users to establish a taxonomic relationship among complex metagenomic datasets. Available from http://www.gobics.de/fabian/treephyler.php (Schreiber et al. 2010).

6.3   Tools for Functional Profiling

(i) *HUMAnN:* HUMAnN predicts the presence or absence along with abundance of microbial pathways in meta-omics data. HUMAnN offers an accurate and efficient approach for characterization of metabolic pathways and functional modules of microbial origin in sequencing reads, allowing the assessment of community roles in metagenomic studies (Abubucker et al. 2012).

(ii) *metaSHARK:* metaSHARK, a "metabolic search and reconstruction kit," is a database and facilitates users with a completely interactive method to explore the KEGG metabolic network through a www browser. It can be found online at http://bioinformatics.leeds.ac.uk/shark/ (Hyland et al. 2006).

(iii) *PRMT:* Predicted-Relative-Metabolic-Turnover, abbreviated as PRMT, allows exploration of metabolite-space derived from the metagenome and can be used as a common tool for the analysis of massive DNA sequences or transcriptome metadata (Larsen et al. 2011).

(iv) *RAMMCAP:* RAMMCAP, "Rapid-Analysis of Multiple-Metagenome," provides a better platform for users to cluster and annotate the metagenome of interest. It was developed on the basis of an ultra-speed processing, fine statistical package and exclusive graphic user interface (GUI). Tool is accessible from http://tools.camera.calit2.net/camera/rammcap/ (Li 2009).

6.4   Tools for Underlying Interactome

(i) *Spar-CC:* A command-based inference program that allows users to perform various important analyses related to the correlation, bootstraps value, and p-values to find the microbial intercommunications network. Available from https://bitbucket.org/yonatanf/sparcc (Friedman and Alm 2012).

(ii) *CCREPE:* CCREPE can be applied for the prediction of microbiome interactions within and between distinct environments and determine their interrelationships under the complex network (Faust et al. 2012)

6.5   Tools for Statistical Tests

(i) *Metastats*: Metastats was the foremost statistical approach developed specially to handle questions in medical research. This tool permits the study of parallel metagenomic reads from two different populations simultaneously and recognizes those characteristics that statistically differentiate the two populations (Paulson et al. 2011).

(ii) *LefSe:* LefSe is an approach for genetic biomarker research through distinct categorization, via their matching and evaluation-consistent biological response. It is accessible from http://huttenhower.sph.harvard.edu/lefse/ (Segata et al. 2011).

(iii) *ShotgunFunctionalizeR:* ShotgunFunctionalizeR is a tool for functional comparative assessment of genetically identified reads. This program includes methods for import, annotation, and visualization of meta-genomics dataset of shotgun high-throughput sequencing. It is available at http://shotgun.zool.gu.se

(iv) *SourceTracker:* SourceTracker is a Bayesian method for estimation of the fraction of a novel community from a group of source environments. It is able to find the rate of invasion or growth of one community among others in any particular set of microbiomes (Knights et al. 2011).

6.6 Simulators Tools

(i) *GemSIM:* General-Error-Model based SIMulators are high-throughput sequencing simulation predictors with the ability of producing the single or paired-end contigs for any sequencing technology compatible with the generic formats SAM and FASTQ (McElroy et al. 2012).

(ii) *MetaSim:* MetaSim is a gene sequence simulation model and can be applied to create groups of synthetic reads that reveal the phylogenetic relationship of the usual metagenome (Richter et al. 2008).

6.7 Tools for Single-Cell Sequencing Analysis

(i) *IDBA-UD:* IDBA-UD is a tool for the assembly of the poor coverage data of the single-cell or metagenome reads with unequal depths. This tool is based on the de Bruijn graph and can be found on http://www.cs.hku.hk/~alse/idba_ud (Peng et al. 2012).

(ii) *SmashCell:* SmashCell is a automated software framework planned to perform the microbial genomic processing, that is, assembly, gene prediction, and functional characterization in a manner that facilitates parameter and algorithm assessment at every step in the process. The source code and manual of SmashCell are available at http://asiago.stanford.edu/SmashCell (Harrington et al. 2010)

6.8   General Toolkits

   (i) *CAMERA:* CAMERA provides users with a rich, unique data bank and bioinformatics tools collection for metagenomics reads processing and allows researchers to disclose the biology of environmental microorganisms. Online website is for the same is http://camera.calit2.net (Seshadri et al. 2007)

  (ii) *IMG/M*: Integrated Microbial Genomes and Metagenome allows users to do comparative analysis of a soil microbiome. It is online accessible at http://img.jgi.doe.gov/m (Markowitz et al. 2012).

 (iii) *MEGAN:* MEGAN is a widely utilized tool for the analysis of metagenomics; rRNA reads along with the meta-transcriptome and metaproteome datasets (Huson et al. 2007).

 (iv) *METAREP:* METAREP or JCVI Metagenomics Reports is a web application for the analysis and comparison of annotated metagenomics reads. METAREP is highly suitable for performing the complete analysis from diversity structure to the metabolome. The website of METAREP is http://www.jcvi.org/metarep (Goll et al. 2010)

  (v) *MG-RAST:* MG-RAST is a high-throughput pipeline for high-performance computing and analysis of metagenomics datasets. The pipeline allows automated functional assignments of metagenomic datasets by comparison with already deposited protein and nucleotide banks. Evolutionary and functional attributes can also be identified. MG-RAST is available at http://metagenomics.nmpdr.org or http://metagenomics.anl.gov. MG-RAST is a fast-growing and constantly updating tool (Meyer et al. 2008; Wilke et al. 2016).

 (vi) *SmashCommunity*: Smash-Community is a self-sufficient annotation tool and provides a unique platform appropriate for analysing the datasets from Sanger and 454 sequencer. Source code and manual of SmashCommunity are available at http://www.bork.embl.de/software/smash (Arumugam et al. 2010).

(vii) *STAMP:* STAMP is a graphical tool package that facilitates mathematical hypothesis analysis and plots for analysis of phylogenetic along with functional profiles. The Python source code and binaries of STAMP can be downloaded from http://kiwi.cs.dal.ca/Software/STAMP (Parks et al. 2014).

(viii) *VAMPS:* VAMPS, the "Visualization and Analysis of Microbial Population Structures," assists researchers working on projects with large-scale sequencing data. A VAMP permits researchers via marker gene sequence data for analysis of microbial community diversity and the interconnection among communities. VAMPS is online accessible from http://vamps.mbl.edu (Huse et al. 2014).

 (ix) *EBI-Metagenomics:* EBI metagenomics allows researchers to process raw genomic reads for function-based attributes analysis and phylogenetic characterization without any difficulty and provides an efficient platform and further automatically stores the data in the European Nucleotide Archive with the associated metadata. EBI metagenomics is online accessible from http://www.ebi.ac.uk/metagenomics/ (Hunter et al. 2014a, b).

Chapter 7
Conclusion and Future Perspectives

Abstract Harnessing the potential of the microbiome can provide cleaner and greener solutions to various environmental challenges. Globally, microbiome research is a continuously evolving field, with researchers aiming to explore the structural and functional characteristics of the microbiomes in various environmental compartments. Still, there are various system-level knowledge gaps in microbiome research. Therefore, unravelling the complex and dynamic microbial crosstalk in the soil system and the multitrophic interactions will certainly provide new impetus for harnessing the real potential of the soil microbiome for multipurpose environmental benefits.

Keywords Environmental challenges · Functional microbiome · Knowledge gap · Multitrophic interactions

7.1 Conclusion

A major part of underground microbial life needs exploration, identification, and characterization. Soil microbial communities are important assets for all type of ecosystems that help in myriads of soil processes such as biogeochemical cycling, nutrient turnover, soil respiration, and various other soil–atmosphere feedbacks by regulating its community structure, functional attributes, substrate specificity, and hormonal and enzymatic profile (Dubey et al. 2016b). The warming climate also presents many aboveground and belowground changes in soil organic matter, root exudation, microbial community structure, pest- and pathogen-mediated disease incidence, plant species diversity, and distribution, etc. Therefore, in-depth knowledge about the structure and functions of the microbial community and its underlying processes is needed now and will certainly provide sustainable solutions for the global food, feed, fiber, and energy demands of the rapidly growing population under changing climatic conditions. This huge microbial world has an enormous role in sustaining plant and human life on planet Earth through improving soil quality, plant disease tolerance, agricultural yield and nutritional quality, restoration

© The Author(s), under exclusive license to Springer Nature Switzerland AG 2020
R. K. Dubey et al., *Unravelling the Soil Microbiome*, SpringerBriefs in
Environmental Science, https://doi.org/10.1007/978-3-030-15516-2_7

potential of degraded lands, organic farming, sustainable phyto-bioremediation, carbon sequestration, and biomass for bioenergy production (Abhilash et al. 2016b).

This book demonstrates that "global soils and agro-ecosystems are under threat" because current agricultural practices such as intensification, extensification, excessive use of agrochemicals, land use patterns, and greenhouse gas (GHG) emissions have exerted tremendous pressure on agro-ecosystems, leading to environmental instability. To consider environmental instability and improve the socioeconomic status of marginal and resource-poor agricultural farmers and practitioners, for sustainable food, fiber, and fuel production and the restoration of degraded land, the exploration of soil and plant and earth microbiome will provide multipurpose benefits to soil, water, air, and human health and environmental sustainability. In approaches promoting the inoculation of plant growth-promoting bacteria and fungi, microbial endophytes, arbuscular mycorrhizal fungi (AMF), and organic amendments, sustainable agronomic practices could be utilized for increasing sustainable food production and agro-ecosystems services (Abhilash et al. 2016a). It would be a better approach if the percentage of arable land area can be increased as the result of degraded land restoration via advanced and sustainable phyto-bioremediation, which generates a huge bioeconomy with minimum environmental risks.

Furthermore, microbiomes act as a hub for myriads of soil system processes and services, the plant immune system and metabolism, various signatory molecules such as microbial volatile organic compounds (mVOCs), genes, and microbe–microbe, plant–plant, and plant–microbe interactions which thus determine plant microbial community composition, plant health, and productivity. However, the aforementioned processes and different plant spheres (spermosphere, rhizosphere, phyllosphere, endosphere), which are supposed to be microbiologically active and dynamic zones for microbial cross-talk, are still poorly explored in terms of microbiome structure and functions. So far, only rare metagenomic identification and comparison of microbial communities among these plant spheres have been done, to the best of our knowledge. The biosynthetic process of microbial VOCs, its genetic regulation, and effect on other microbial partners from different plant spheres remains unclear. In such a scenario, application of conventional, as well as the next-generation technologies (metagenomics, proteomics, transcriptomics), for studying the whole earth microbiome and its associated ecosystem services is an urgent need. Unravelling the global microbiomes will surely open unexplored treasures for improving soil system and environmental sustainability and human well-being and will be helpful in achieving the sustainable development goals (SDGs), via attaining its targets: 1 (no poverty), 2 (zero hunger), 3 (good health and well-being), 6 (clean water and sanitation), 7 (affordable and clean energy, and 12 (responsible consumption and productions), 13 (climate action), 14 (life below water), and 15 (life on land).

7.2   Future Microbiome Research Directions: How Do They Engage Themselves?

As we explained earlier, microorganisms living in close proximity to crop plants interact with them in many ways. Although the interaction is complex and has a strong evolutionary linkage, such partnerships can have long-term impacts on both crop plants and their respective microbial partners (Badri and Vivanco 2009; Tewari and Arora 2013). The root systems act as a chemical repository for releasing specialized biomolecules to interact with numerous beneficial microorganisms such as rhizobia, mycorrhizae, endophytes, and other plant growth-promoting microbes (PGPMs) in the soil. As a result, a particular plant species could keep a highly specific microbial diversity in their rhizobiome, mainly through the modulation of cross-talk between their roots and rhizospheric microbes through the mediation of specific root exudates (Huang et al. 2014). Moreover, the exudation of a variety of low molecular weight organic compounds (e.g., sugars, polysaccharides, amino acids, organic acids, phenolic compounds), high molecular weight organic compounds (e.g., mucilage and proteins), and even volatile organic compounds (VOCs; e.g., CO_2, alcohols, aldehydes, terpenes) into the soil system would modify its physical, chemical, and biochemical properties by altering the nutrient stature, pH-redox modulating factors, acidity, alkalinity, moisture content, porosity, permeability, particle density, and bulk density of the soil aggregates (Tewari and Arora 2013). Thus, a plant species can customize its rhizobiome for the preferred microbial partners (Berendsen et al. 2012). On the basis of the selectivity, specificity, and recognition of various signaling cues, microorganisms can establish a precise association with a host plant. In return, microorganisms produce various phytohormones such as auxins, cytokinins, gibberellins, abscisic acid, and polyamines for the growth and proliferation of the plant. Furthermore, most of the microbial volatiles have a key role in eliciting induced systemic resistance (ISR) in plants. Apart from this, the PGPMs reduce ethylene stress in plants by the secretion of ACC deaminase and are involved in nutrient availability to plants by biological nitrogen fixation, phosphate, potassium, and zinc solubilization, and ion chelation by siderophore production (Ghosh et al. 2014; Nadeem et al. 2014). Above all, microbial signals to plants improve the cell metabolism, growth, development, and productivity. Because of the ecological and health hazards of agrochemicals, soil and plant–microbiome research can be adjudged as a sustainable solution for improving global agricultural productivity, soil management, and ecosystem services that surely will meet the target of the United Nations Sustainable Development Goals (SDGs). The aforesaid approaches and convention explore the multifaceted applications of plant–microbe partnerships for sustaining food production and plant, human, and environmental health. In such a scenario, application of conventional methods, as well as the next-generation technologies (metagenomics, proteomics, transcriptomics), for studying the whole earth microbiome and its associated ecosystem services is an urgent need. Recent global statistics suggested that microbial diversity analyses across multiple research areas are shifting to the use of next-generation sequencing (NGS) technologies (Fig. 7.1).

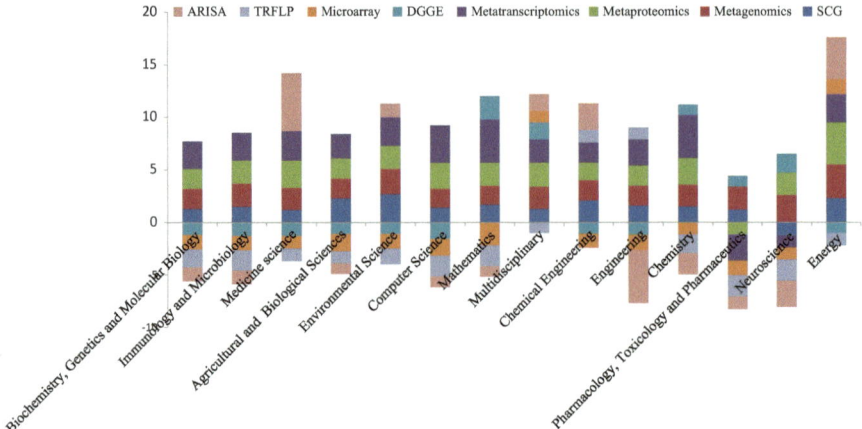

Fig. 7.1 Global statistics showing recent trends of major techniques and approaches utilized for microbial diversity analysis. In a recent trend, single-cell genomics, metagenomics, metaproteomics, and metatranscriptomics are the techniques most applied, whereas microarray, deferential gradient gel electrophoresis (DGGE), temperature restriction fragment length polymorphism (TRFLP), and automated ribosomal intergenic spacer analysis (ARISA) techniques have been utilized in earlier studies but not in recent trends. The scenario of agricultural and biological sciences has highlighted similar trends with a 2.3-fold increase in single-cell genomics and metatranscriptomics as well as a 1.9-fold increase in the application of metagenomics and metaproteomics. Similar trends of microbiome studies in environmental sciences have also been found where statistics suggested a 2.7-, 2.4-, 2.2-, and 2.7-fold increase in the application of single-cell genomics, metagenomics, metaproteomics, and metatranscriptomics, respectively

To save time, obtain precise results, and improve our understanding about microbial community structure and functions, an integrated methodology may also be adopted.

Nowadays, the microbiome is a burgeoning area of research under all scientific disciplines such as plant, animal, and medical science. The microbiome is ubiquitous in nature and is involved in the various underlying mechanisms of life, regulating the health of soil, plants, and humans under normal and extreme environments. Furthermore, certain innovations in the microbial world are indicating new directions that lead to the huge complexity under microbiome research. To obtain a clear picture of the microbiome, we have to keep these points under consideration when conducting any microbiome study. (i) Recently, microorganisms such as *Paenibacillus* sp. have been reported to establish a negative interaction and change the environmental pH to an extent that is lethal to other microbial communities in similar habitats (Ratzke et al. 2018). This phenomenon, known as ecological suicide, has a major role in the structure, evolution, interaction, and functions of microbial communities under diverse ecosystems. (ii) The microbial community has an intertwined microbial network retaining keystone taxa that act as the driver of microbiome structure and function (Banerjee et al. 2018). For example, agricultural soils have various keystone taxa belonging to *Gemmatimonas*, Acidobacteria GP17, Xanthomonadales, Rhizobiales, Burkholderiales, Solirubrobacterales, and Verrucomicrobia, which regulate microbial communities and the dynamics of many agroecosystem processes. So, the presence or speciation of such keystone taxa or microorganisms causing ecological suicide may certainly alter microbial community structure and functions. (iii) The third point is related to the technological aspect of microbiome analysis. The operational taxonomic unit (OTU) has been much explored to define the structure of the microbiome, but recent studies suggest going beyond the OTU and emphasizing the use of "exact sequence variant" for microbiome analysis (Knight et al. 2018). These aforesaid points should be considered during all microbiome analysis. The pan-genome can also be utilized while studying the microbial community structure of any habitat. Pan-genome is a metagenomic tool to provide better understanding of the phylogenetic resolution of the microbial genome. Pan-genome covers the extended core genes (control translation, replication), characters gene (control photosynthesis, endosymbionts, adaptation to environmental niche), and accessory gene pool of a microbial species. Greater niche diversity has a larger pan-genome size and has a greater opportunity for lateral gene flux across the strain, providing the framework for understanding the mechanism of species evolution and estimating genomic diversity (Bentley 2009). Limited dataset, inadequate functional features, and difficult installation are the major problems of existing pan-genome software. This problem can be resolved using the ultrafast computational pipeline bacterial pan-genome analysis tool (BPGA). BPGA combined with a diverse clustering method such as USEARCH, CD-HIT, or OrthoMCL provide novel features for downstream analysis such as multi-locus sequence typing (MLST) phylogeny, presence/absence of genes in the specific strain, unique genes, and atypical G+C content analysis (Chaudhari et al. 2016). Advanced DNA sequencing, metagenomic, and metabolomic technology are producing large datasets in the field of microbiome research. These datasets need precise and coherent analysis to acquire correct information to explore bacterial community composition, interactions, and their role in plant, human, and environmental health and ecosystem services.

Navigating the whole earth microbiome, oceanic microbiome, seed microbiome, plant microbiome, and human gut microbiome are ongoing research efforts exploring the potential of the complete microbiome in ocean science, soil ecological science, agricultural science, and human health (Turnbaugh et al. 2007; Tseng and Tang 2014; Gilbert et al. 2014; Berg and Raaijmakers 2018). Knowledge acquired from these studies must rectify and enhance scientific understanding and help us meet the global challenges for developing a sustainable world.

References

Abdeljailil NO-B, Renault D, Gerbore J, Vallance J, Rey P, Daami-Remadi M (2016) Evaluation of the effectiveness of tomato-associated rhizobacter applied singly or as three-strain consortium for biosuppression of *Sclerotinia* stem rot in tomato. J Microb Biochem Technol 8:4

Abhilash PC, Dubey RK (2014) Integrating aboveground–belowground responses to climate change. Curr Sci 1637:12

Abhilash PC, Dubey RK (2015) Root system engineering: prospects and promises. Trends Plant Sci 20:408–409

Abhilash PC, Yunus M (2011) Can we use biomass produced from phytoremediation? Biomass Bioenergy 35:1371–1372

Abhilash PC, Srivastava P, Jamil S, Singh N (2011) Revisited *Jatropha curcas* as an oil plant of multiple benefits: critical research needs and prospects for the future. Environ Sci Pollut Res 18:127–131

Abhilash PC, Powell JR, Singh HB, Singh BK (2012) Plant–microbe interactions: novel applications for exploitation in multipurpose remediation technologies. Trends Biotechnol 30:416–420

Abhilash PC, Dubey RK, Tripathi V, Srivastava P, Verma JP, Singh HB (2013a) Adaptive soil management. Curr Sci 104:1275–1276

Abhilash PC, Dubey RK, Tripathi V, Srivastava P, Verma JP, Singh HB (2013b) Remediation and management of POPs-contaminated soils in a warming climate: challenges and perspectives. Environ Sci Pollut Res 20:5879–5885

Abhilash PC, Tripathi V, Dubey RK, Edrisi SA (2015) Coping with changes: adaptation of trees in a changing environment. Trends Plant Sci 20:137–138

Abhilash PC, Dubey RK, Tripathi V, Gupta VK, Singh HB (2016a) Plant growth-promoting microorganisms for environmental sustainability. Trends Biotechnol 34:847–850

Abhilash PC, Tripathi V, Edrisi SA, Dubey RK, Bakshi M, Dube PK, Ebbs SD (2016b) Sustainability of crop production from polluted lands. Energy Ecol Environ 1:54–65

Abubucker S, Segata N, Goll J, Schubert AM, Izard J, Cantarel BL et al (2012) Metabolic reconstruction for metagenomic data and its application to the human microbiome. PLoS Comput Biol 8:e1002358

Abulencia CB, Wyborski DL, Garcia JA, Podar M, Chen W, Chang SH, Chang HW, Watson D, Brodie EL, Hazen TC, Keller M (2006) Environmental whole-genome amplification to access microbial populations in contaminated sediments. Appl Environ Microbiol 72:3291–3301

Adams P, Lynch JM, De Leij FAAM (2007) Desorption of zinc by extracellularly produced metabolites of *Trichoderma harzianum*, *Trichoderma reesei* and *Coriolus versicolor*. J Appl Microbiol 103:2240–2247

© The Author(s), under exclusive license to Springer Nature Switzerland AG 2020
R. K. Dubey et al., *Unravelling the Soil Microbiome*, SpringerBriefs in Environmental Science, https://doi.org/10.1007/978-3-030-15516-2

Adesemoye AO, Kloepper JW (2009) Plant–microbes interactions in enhanced fertilizer-use efficiency. Appl Microbiol Biotechnol 85:1–12

Aguiar-Pulido V, Huang W, Suarez-Ulloa V, Cickovski T, Mathee K, Narasimhan G (2016) Metagenomics, metatranscriptomics, and metabolomics approaches for microbiome analysis: supplementary issue: bioinformatics methods and applications for big metagenomics data. Evol Bioinform 12:EBO-S36436

Akinsemolu AA (2018) The role of microorganisms in achieving the sustainable development Goals. J Clean Prod 182:139–155

Albantoglu OU, Way SF, Hinrichs SH, Sayood K (2011) RAIphy: phylogenetic classification of metagenomics samples using iterative refinement of relative abundance index profiles. BMC Bioinform 12:41

Amann R, Fuchs BM (2008) Single-cell identification in microbial communities by improved fluorescence in situ hybridization techniques. Nat Rev Microbiol 6:339–348

Amico SD, Collins T, Marx J-C, Feller G, Gerday C (2006) Psychrophilic microorganisms: challenges for life. EMBO Rep 7:385–389

Anderson LB, Maderia M, Ouellette AJA, Putman-Evans C, Higgins L, Krick T, MacCoss MJ, Lim H, Yates JR, Barry BA (2002) Post translational modifications in the CP43 subunit of photosystem II. Proc Natl Acad Sci U S A 23:14676–14681

Andreolli M, Lampis S, Poli M, Gullner G, Biró B, Vallini G (2013) Endophytic Burkholderia fungorum DBT1 can improve phytoremediation efficiency of polycyclic aromatic hydrocarbons. Chemosphere 92:688–694

Anupama NB, Jogaiah S, Ito S, Amruthesh KN, Tran LP (2015) Improvement of growth, fruit weight and early blight disease protection of tomato plants by rhizosphere bacteria is correlated with their beneficial traits and induced biosynthesis of antioxidant peroxidase and polyphenol oxidase. Plant Sci 231:62–73

Arumugam M, Harrington ED, Foerstner KU, Raes J, Bork P (2010) Smash Community: a metagenomic annotation and analysis tool. Bioinformatics 26:2977–2978

Aziz RK, Bartels D, Best AA, DeJongh M, Disz T, Edwards RA, Meyer F et al (2008) The RAST server: rapid annotations using subsystems technology. BMC Genomics 9(1):75. https://doi.org/10.1186/1471-2164-9-75

Babu AG, Shim J, Shea PJ, Oh BT (2014a) *Penicillium aculeatum* PDR-4 and *Trichoderma* sp. PDR-16 promote phytoremediation of mine tailing soil and bioenergy production with sorghum-sudan grass. Ecol Eng 69:186–191

Babu AG, Shim J, Bang KS, Shea PJ, Oh BT (2014b) *Trichoderma virens* PDR-28: a heavy metal-tolerant and plant growth-promoting fungus for remediation and bioenergy crop production on mine tailing soil. J Environ Manag 132:129–134

Badri DV, Vivanco JM (2009) Regulation and function of root exudates. Plant Cell Environ 32:666–681

Bahadur A, Ahmad R, Afzal A, Feng H, Suthar V, Batool A, Mahmood-ul-Hassan M (2017) The influences of Cr-tolerant rhizobacteria in phytoremediation and attenuation of Cr (VI) stress in agronomic sunflower (*Helianthus annuus* L.). Chemosphere 179:112–119

Bai Y, Liang J, Liu R, Hu C, Qu J (2014) Metagenomic analysis reveals microbial diversity and function in the rhizosphere soil of a constructed wetland. Environ Technol 35:2521–2527

Balseiro-Romero M, Gkorezis P, Kidd PS, Van Hamme J, Weyens N, Monterroso C, Vangronsveld J (2017) Use of plant growth promoting bacterial strains to improve *Cytisus striatus* and *Lupinus luteus* development for potential application in phytoremediation. Sci Total Environ 581:676–688

Banerjee S, Schlaeppi K, Heijden MG (2018) Keystone taxa as drivers of microbiome structure and functioning. Nat Rev Microbiol. https://doi.org/10.1038/s41579-018-0024-1

Baraniya D, Puglisi E, Ceccherini MT, Pietramellara G, Giagnoni L, Arenella M, Renella G (2016) Protease encoding microbial communities and protease activity of the rhizosphere and bulk soils of two maize lines with different N uptake efficiency. Soil Biol Biochem 96:176–179

Bashan LE, Hernandez JP, Bashan Y (2012) The potential contribution of plant growth-promoting bacteria to reduce environmental degradation – a comprehensive evaluation. Appl Soil Ecol 61:171–189

Bashiardes S, Shapiro H, Rozin S, Shibolet O, Elinav E (2016a) Non-alcoholic fatty liver and the gut microbiota. Mol Metab 5:782–794

Bashiardes S, Zilberman-Schapira G, Elinav E (2016b) Use of metatranscriptomics in microbiome research. Bioinform Biol Insights 10:BBI-S34610. https://doi.org/10.4137/BBI.S34610. eCollection 2016

Bashir Y, Singh SP, Konwar BK (2014) Metagenomics: an application based perspective. Chin J Biol. https://doi.org/10.1155/2014/146030

Bastida F, Moreno JL, Nicolas C, Hernandez T, Garcia C (2009) Soil metaproteomics: a review of an emerging environmental science. significance, methodology and perspectives. Eur J Soil Sci 60:845–859

Beauregard PB, Chai Y, Vlamakis H, Losick R, Kolter R (2013) Bacillus subtilis biofilm induction by plant polysaccharides. Proc Natl Acad Sci U S A 110:E1621–E1630

Becker JM, Parkin T, Nakatsu CH, Wilbur JD, Konopka A (2006) Bacterial activity, community structure and centimeter-scale spatial heterogeneity in contaminated soils. Microbiol Ecol 51:220–231

Begum MM, Sariah M, Puteh B, Zainal Abidin M, Rahman M, Siddiqui Y (2010) Field performance of bio-primed seeds to suppress *Colletotrichum truncatum* causing damping-off and seedling stand of soybean. Biol Control 53:18–23

Ben-David EA, Holden PJ, Stone DJ, Harch BD, Foster LJ (2004) The use of phospholipid fatty acid analysis to measure impact of acid rock drainage on microbial communities in sediments. Microb Ecol 48:300–315

Ben-David EA, Zaady E, Sher Y, Nejidat A (2011) Assessment of the spatial distribution of soil microbial communities in patchy arid and semi-arid landscapes of the Negev Desert using combined PLFA and DGGE analyses. FEMS Microbiol Ecol 76:492–450

Benndorf D, Balcke GU, Harms H, von Bergen M (2007) Functional metaproteome analysis of protein extracts from contaminated soil and groundwater. ISME J 1:224–234

Bentley S (2009) Sequencing the species pan-genome. https://doi.org/10.1038/nrmicro2123

Berendsen R, Pieterse C, Bakker P (2012) The rhizosphere microbiome and plant health. Trends Plant Sci 17:478–486

Berg G, Raaijmakers JM (2018) Saving seed microbiomes. ISME J 12:1167–1170

Berger SA, Stamatakis A (2011) Aligning short reads to reference alignments and trees. Bioinformatics 27:2068–2075

Berta G, Copetta A, Gamalero E, Bona E, Cesaro P, Scarafoni A, D'Agostino G (2014) Maize development and grain quality are differentially affected by mycorrhizal fungi and a growth-promoting pseudomonad in the field. Mycorrhiza 24:161–170

Bertani I, Abbruscato P, Piffanelli P, Subramoni S, Venturi V (2016) Rice bacterial endophytes: isolation of a collection, identification of beneficial strains and microbiome analysis. https://doi.org/10.1111/1758-2229.12403

Beverley SM et al (2002) Putting the *Leishmania* genome to work: functional genomics by transposon trapping and expression profiling. Philos Trans R Soc Lond B 357:47–53

Blainey PC (2012) The future is now: single-cell genomics of bacteria and archaea. https://doi.org/10.1111/1574-6976.12015

Bodrossy L, Sessitsch A (2004) Oligonucleotide microarrays in microbial diagnostics. Curr Opin Microbiol 7:245–254

Bogan BW, Lamar RT (1996) Polycyclic aromatic hydrocarbon-degrading capabilities of *Phanerochaete laevis* HHB-1625 and its extracellular ligninolytic enzymes. Appl Environ Microbiol 62:1597–1603

Bolger AM, Lohse M, Usadel B (2014) Trimmomatic: a flexible trimmer for Illumina sequence data. Bioinformatics 30:2114–2120

Brady A, Salzberg SL (2009) Phymm and Phymm BL: metagenomic phylogenetic classification with interpolated Markov models. Nat Methods 6:673–676

Brady A, Salzberg S (2011) PhymmBL expanded: confidence scores, custom databases, parallelization and more. Nat Methods 8(5):367. https://doi.org/10.1038/nmeth0511-367

Braud A, Jézéquel K, Bazot S, Lebeau T (2009) Enhanced phytoextraction of an agricultural Cr-and Pb-contaminated soil by bioaugmentation with siderophore-producing bacteria. Chemosphere 74:280–286

Brown MV, Fuhrman JA (2005) Marine bacterial microdiversity as revealed by internal transcribed spacer analysis. Aquat Microb Ecol 41:15–23

Bulgarelli D et al (2013) Structure and functions of the bacterial micro biota of plants. Annu Rev Plant Biol 64:807–838

Burd GI, Dixon DG, Glick BR (2000) Plant growth-promoting bacteria that decrease heavy metal toxicity in plants. Can J Microbiol 46:237–245

Butterly CR, Phillips LA, Wiltshire JL, Franksc AE, Armstronga RD, Chene D, Melea PM, Tanga C (2016) Long-term effects of elevated CO2 on carbon and nitrogen functional capacity of microbial communities in three contrasting soils. Soil Biol Biochem 97:157–167

Calderon FJ, Nielsen ID, Acosta-Martinez V, Vigil MF, Lyon D (2016) Cover crop and irrigation effects on soil microbial communities and enzymes in semiarid agroecosystems of the central Great Plains of North America. Pedosphere 26:192–205

Canfield DE, Glazer AN, Falkowski PG (2010) The evolution and future of Earth's nitrogen cycle. Science 330:192–196

Caracciolo AB, Bottoni P, Grenni P (2010) Fluorescence in situ hybridization in soil and water ecosystems: a useful method for studying the effect of xenobiotics on bacterial community structure. Toxicol Environ Chem 92:567–579

Carr R, Borenstein E (2014) Comparative analysis of functional metagenomic annotation and the mappability of short reads. PLoS One 9:e105776

Carvalhais LC, Dennis PG, Tyson GW, Schenk PM (2012) Application of metatranscriptomics to soil environments. J Microbiol Methods 91:246–251

Chang P, Gerhardt KE, Huang XD, Yu XM, Glick BR, Gerwing PD, Greenberg BM (2014) Plant growth-promoting bacteria facilitate the growth of barley and oats in salt-impacted soil: implications for phytoremediation of saline soils. Int J Phytoremediation 16:1133–1147

Charuvaka A, Rangwala H (2011) Evaluation of short read metagenomic assembly. BMC Genomics 12:S8

Chaudhari NM, Gupta VK, Dutta C (2016) BPGA-an ultra-fast pan-genome analysis pipeline. Sci Rep. https://doi.org/10.1038/srep24373

Chauhan A, Smartt A, Wang J, Utturkar S, Frank A, Bi M, Arreaza A (2014) Integrated metagenomics and metatranscriptomics analyses of root-associated soil from transgenic switchgrass. Genome Announc 2(4):e00777–e00714. https://doi.org/10.1128/genomeA.00777-14

Chen Y, Murrell JC (2010) When metagenomics meets stable-isotope probing: progress and perspectives. Trends Microbiol 18:157–163

Chen Y, Dumont MG, McNamara NP, Chamberlain PM, Bodrossy L, Pavese NS, Murrell JC (2008) Diversity of the active methanotrophic community in acidic peatlands as assessed by mRNA and SIP-PLFA analyses. Environ Microbiol 110:446–459

Chen LX, Hu M, Huang LN, Hua ZS, Kuang JL, Li SJ, Shu WS (2015) Comparative metagenomic and metatranscriptomic analyses of microbial communities in acid mine drainage. ISME J 9(7):1579–1592

Chen IMA, Markowitz VM, Chu K, Palaniappan K, Szeto E, Pillay M, Ratner A, Huang J, Andersen E, Huntemann M, Varghese N, Hadjithomas M, Tennessen K, Nielsen T, Ivanova NN, Kyrpides NC (2017) IMG/M: integrated genome and metagenome comparative data analysis system. Nucleic Acids Res 4:D507–D516

Choksawangkarn W, Edwards N, Wang Y, Gutierrez P, Fenselau C (2012) Comparative study of workflows optimized for in-gel, in-solution, and on-filter proteolysis in the analysis of plasma membrane proteins. J Proteome Res 11:3030–3034

Chourey K, Jansson J, VerBerkmoes N, Shah M, Chavarria KL, Tom LM, Brodie EL, Hettich RL (2010) Direct cellular lysis/protein extraction protocol for soil metaproteomics. J Proteome Res 9:6615–6622

Cole JR, Wang Q, Cardenas E, Fish J, Chai B, Farris RJ, Kulam-Syed-Mohideen AS, McGarrell DM, Marsh T, Garrity GM, Tiedje JM (2009) The Ribosomal Database Project: improved alignments and new tools for rRNA analysis. Nucleic Acids Res 37:D141–D145

Collins G, Kavanagh S, Mchugh S, Connaughton S, Kearney A, Rice O, Carrigg C, Scully C, Bhreathnach N, Mahony T, Madden P, Enright AM, Flaherty V (2006) Accessing the black box of microbial diversity and ecophysiology: recent advances through polyphasic experiments. J Environ Sci Health A 41:897–922

Conesa A, Götz S, García-Gómez JM, Terol J, Talón M, Robles M (2005) Blast2GO: a universal tool for annotation, visualization and analysis in functional genomics research. Bioinformatics 21:3674–3676

Cong J, Yang Y, Liu X, Lu H, Liu X, Zhou J, Li D, Yin H, Ding J, Zhang Y (2015) Analyses of soil microbial community compositions and functional genes reveal potential consequences of natural forest succession. Sci Rep-UK 5:10007

Costa R, Götz M, Mrotzek N, Lottmann J, Berg G, Smalla K (2006) Effects of site and plant species on rhizosphere community structure as revealed by molecular analysis of different microbial guilds. FEMS Microbiol Ecol 56:236–249

Couillerot O, Ramírez-Trujillo A, Walker V, von Felten A, Jansa J, Maurhofer M, Moënne-Loccoz Y (2013) Comparison of prominent Azospirillum strains in Azospirillum-Pseudomonas-Glomus consortia for promotion of maize growth. Appl Microbiol Biotechnol 97:4639–4649

Creer S, Deiner K, Frey S, Porazinska D, Taberlet P, Thomas WK, Potter C, Bik HM (2016) The ecologist's field guide to sequence-based identification of biodiversity. Methods Ecol Evol 7:1008–1018

Damon C, Lehembre F, Oger-Desfeux C, Luis P, Ranger J, Fraissinet-Tachet L, Marmeisse R (2012) Metatranscriptomics reveals the diversity of genes expressed by eukaryotes in forest soils. PLoS One 7(1):e28967. https://doi.org/10.1371/journal.pone.0028967

Dang QL, Shin TS, Park MS, Choi YH, Choi GJ, Jang KS, Kim IS, Kim J-C (2014) Antimicrobial activities of novel mannosyl lipids isolated from the biocontrol fungus Simplicillium lamellicolaBCP against Phytopathogenic Bacteria. J Agric Food Chem 62:3363–3370

Darling AE, Jospin G, Lowe E, Matsen FA, Bik HM, Eisen JA (2014) Phylogenetic analysis of genomes and metagenomes. PeerJ 2:e243

De Bourcy CF, De Vlaminck I, Kanbar JN, Wang J, Gawad C, Quake SR (2014) A quantitative comparison of singlecell whole genome amplification methods. PLoS One 9(8):e105585. https://doi.org/10.1371/journal.pone.0105585

de la Luz Mora M, Demaneta R, Acunaa JJ, Viscardia S, Jorqueraa M, Rengelb Z, Duran P (2017) Aluminium-tolerant bacteria improve the plant growth and phosphorus content in ryegrass grown in a volcanic soil amended with cattle dung manure. Appl Soil Ecol 115:19–26

de-Bashan LE, Hernandez JP, Bashan Y (2012) The potential contribution of plant growth-promoting bacteria to reduce environmental degradation – a comprehensive evaluation. Appl Soil Ecol 61:171–189

Degefu Y, Somervuo P, Aittamaa M, Virtanen E, Valkonen JPT (2016) Evaluation of a diagnostic microarray for the detection of major bacterial pathogens of potato from tuber samples. Bull OEPP 46:103–111

Delahunty CM, Yates JR (2007) MudPIT: multidimensional protein identification technology. BioTechniques 43:563–567

Deleye L, Tilleman L, Vander Plaetsen AS, Cornelis S, Deforce D, Van Nieuwerburgh F (2017) Performance of four modern whole genome amplification methods for copy number variant detection in single cells. Sci Rep 7(1):3422

DeLong EF, Wickham GS, Pace NR (1989) Phylogenetic stains: ribosomal RNA-based probes for the identification of single cells. Science 243:1360–1363

Deng Z, Zhang R, Shi Y, Tan H, Cao L (2014) Characterization of Cd-, Pb-, Zn-resistant endophytic Lasiodiplodia sp. MXSF31 from metal accumulating Portulaca oleracea and its potential in promoting the growth of rape in metal-contaminated soils. Environ Sci Pollut Res 21:2346–2357

DeSantis TZ, Hugenholtz P, Larsen N, Rojas M, Brodie EL, Keller K, Huber T, Dalevi D, Hu P, Andersen GL (2006) Greengenes, a chimera-checked 16S rRNA gene database and workbench compatible with ARB. Appl Environ Microbiol 72:5069–5072

Dhillon V, Li X (2015) Single-cell genome sequencing for viral-host interactions. J Comput Sci Syst Biol 8:160–165

Dhiman SS, Selvaraj C, Li J, Singh R, Zhao X, Kim D, Lee JK (2016) Phytoremediation of metal-contaminated soils by the hyperaccumulator canola (*Brassica napus* L.) and the use of its biomass for ethanol production. Fuel 183:107–114

di Gregorio S, Barbafieri M, Lampis S, Sanangelantoni AM, Tassi E, Vallini G (2006) Combined application of Triton X-100 and *Sinorhizobium* sp. Pb002 inoculum for the improvement of lead phytoextraction by *Brassica juncea* in EDTA amended soil. Chemosphere 63:293–299

Diaz NN, Krause L, Goesmann A, Niehaus K, Nattkemper TW (2009) TACOA: taxonomic classification of environmental genomic fragments using a kernelized nearest neighbor approach. BMC Bioinform 10:56

Drenovsky RE, Steenwerth KL, Jackson LE, Scow KM (2010) Land use and climatic factors structure regional patterns in soil microbial communities. Glob Ecol Biogeogr 19:27–39

Drigo B, Van Veen JA, Kowalchuk GA (2009) Specific rhizosphere bacterial and fungal groups respond differently to elevated atmospheric CO2. ISME J 3:1204–1217

Dröge J, McHardy AC (2012) Taxonomic binning of metagenome samples generated by next-generation sequencing technologies. Brief Bioinform 13:646–655

Dubey RK, Tripathi V, Singh N, Abhilash PC (2014) Phytoextraction and dissipation of lindane by *Spinacia oleracea* L. Ecotoxicol Environ Saf 109:22–26

Dubey RK, Tripathi V, Abhilash PC (2015) Book Review: Principles of plant-microbe interactions: microbes for sustainable agriculture. Front Plant Sci. https://doi.org/10.3389/fpls.2015.00986

Dubey PK, Singh GS, Abhilash PC (2016a) Agriculture in a changing climate. J Clean Prod 113:1046–1047

Dubey RK, Tripathi V, Dubey PK, Singh HB, Abhilash PC (2016b) Exploring rhizospheric interactions for agricultural sustainability: the need of integrative research on multi-trophic interactions. J Clean Prod 115:362–365

Dubey RK, Tripathi V, Edrisi SA, Bakshi M, Dubey PK, Singh A, Verma JP, Singh A, Sarma BK, Raskhit A, Singh DP, Singh HB, Abhilash PC (2017) Role of plant growth promoting microorganisms in sustainable agriculture and environmental remediation. In: Singh HB, Sharma B, Kesawani C (eds) Advances in PGPR research. CABI Press. https://doi.org/10.1079/9781786390325.0000

Dumont MG, Murrell JC (2005) Stable isotope probing-linking microbial identity to function. Nat Rev Microbiol 3:499–504

Edrisi SA, Abhilash PC (2016) Exploring marginal and degraded lands for biomass and bioenergy production: an Indian scenario. Renew Sust Energ Rev 54:1537–1551

Edrisi SA, Dubey RK, Tripathi V, Bakshi M, Srivastava P, Jamil S, Abhilash PC (2015) *Jatropha curcas* L.: a crucified plant waiting for resurgence. Renew Sust Energ Rev 41:855–862

Ekblom R, Wolf JBW (2014) A field guide to whole-genome sequencing, assembly and annotation. Evol Appl 7:1026–1042

El-Howeity MA, Asfour MM (2012) Response of some varieties of canola plant (*Brassica napus* L.) cultivated in a newly reclaimed desert to plant growth promoting rhizobacteria and mineral nitrogen fertilizer. Ann Agric Sci 57:129–136

Emmert-Buck MR et al (1996) Laser capture microdissection. Science 274:998–1001

Faust K, Sathirapongsasuti JF, Izard J, Segata N, Gevers D, Raes J et al (2012) Microbial co-occurrence relationships in the human microbiome. PLoS Comput Biol 8:e1002606

Fernandes JP, Marisa C, Almeida R, Andreotti F, Barros L, Almeida T, Mucha P (2017) Response of microbial communities colonizing salt marsh plants rhizosphere to copper oxide nanoparticles contamination and its implications for phytoremediation processes. Sci Total Environ 581:801–810

Ferrer M, Ruiz A, Lanza F, Haange SB, Oberbach A, Till H, Bargiela R, Campoy C, Segura MT, Richter M, von Bergen M, Seifert J, Suarez A (2013) Microbiota from the distal guts of lean and obese adolescents exhibit partial functional redundancy besides clear differences in community structure. Environ Microbiol 15:211–226

Ferrigo D, Raiola A, Rasera R, Causin R (2014) *Trichoderma harzianum* seed treatment controls *Fusarium verticillioides* colonization and fumonisin contamination in maize under field conditions. Crop Prot 65:51–56

Fic E, Kedracka-Krok S, Jankowska U, Pirog A, Dziedzicka-Wasylewska M (2010) Comparison of protein precipitation methods for various rat brain structures prior to proteomic analysis. Electrophoresis 31:3573–3579

Fierer N, Jackson JA, Vilgalys R, Jackson RB (2005) Assessment of soil microbialcommunity structure by use of taxon-specific quantitative PCR assays. Appl Environ Microbiol 71:4117–4120

Figueroa-Lopez AM, Cordero-Ramirez JD, Martinez-Alvarez JC, Lopez-Meyer M, Lizarraga-Sanchez GJ, Felix-Gastelum R, Castro-Martinez C, Maldonado-Mendoza IE (2016) Rhizospheric bacteria of maize with potential for biocontrol of *Fusarium verticillioides*. Springerplus 5:330

Filippo CD, Ramazzotti M, Fontana P, Cavalieri D (2012) Bioinformatic approaches for functional annotation and pathway inference in metagenomics data. Brief Bioinform 13:696–710

Finn RD, Bateman A, Clements J, Coggill P, Eberhardt RY, Eddy SR, Heger A, Hetherington K, Holm L, Mistry J, Sonnhammer EL, Tate J, Punta M (2014) the protein families database. Nucleic Acids Res 42:D222–D230

Fischer SG, Lerman LS (1979) Length-independent separation of DNA restriction fragments in two dimensional gel electrophoresis. Cell 16:191–200

Food and Agriculture Organization of the United Nations (2013) Climate smart agriculture sourcebook. FAO Rome, Italy

Franzmann PD, Patterson BM, Power TR, Nichols PD, Davis GB (1996) Microbial biomass in a shallow, urban aquifer contaminated with aromatic hydrocarbons: analysis of phospholipid fatty acid content and composition. J Appl Bacteriol 80:617–625

Franzosa EA, Morgan XC, Segata N, Waldron L, Reyes J, Earl AM, Izard J (2014) Relating the metatranscriptome and metagenome of the human gut. Proc Natl Acad Sci 111:E2329–E2338. https://doi.org/10.1073/pnas.1319284111

Friedman J, Alm EJ (2012) Inferring Correlation Networks from Genomic Survey Data. PLoS Comput Biol 8:e1002687

Gao X, Gong Y, Huo Y, Han Q, Kang Z, Huang L (2015) Endophytic *Bacillus subtilis* strain E1R-J is a promising biocontrol agent for wheat powdery mildew. Biomed Res Int. https://doi.org/10.1155/2015/462645

Gawad C, Koh W, Quake SR (2016) Single-cell genome sequencing: current state of the science. Nat Rev Genet 3:175–188

Gebreil AS, Abraham W-R (2016) Diversity and activity of bacterial biofilm communities growing on hexachlorocyclohexane. Water Air Soil Pollut 227:295

Gelfand I, Sahajpal R, Zhang X, Izaurralde RC, Gross KL, Robertson GP (2013) Sustainable bioenergy production from marginal lands in the US Midwest. Nature 493:514–517

Gerlach W, Stoye J (2011) Taxonomic classification of metagenomic shotgun sequences with CARMA3. Nucleic Acids Res 39:e91

Ghosh S, Chan CKK (2016) Analysis of RNA-Seq data using TopHat and Cufflinks. In: Plant bioinformatics. Humana Press, New York, pp 339–361

Ghosh TS, Monzoorul Haque M, Mande SS (2010) DiScRIBinATE: a rapid method for accurate taxonomic classification of metagenomic sequences. BMC Bioinform 7:S14

Ghosh UD, Saha C, Maiti M, Lahiri S, Ghosh S, Seal A et al (2014) Root associated iron oxidizing bacteria increase phosphate nutrition and influence root to shoot partitioning of iron in tolerant plant *Typha angustifolia*. Plant Soil. https://doi.org/10.1007/s11104-014-2085-x

Giagnoni L, Magherini F, Landi L, Taghavi S, van der Lelie D, Puglia M, Bianchi L, Bini L, Nannipieri P, Renella G, Modesti A (2012) Soil solid phases effects on the proteomic analysis of *Cupriavidus metallidurans* CH34. Biol Fertil Soils 48:425–433

Gilbert JA, Meyer F, Bailey MJ (2011) The future of microbial metagenomics (or is ignorance bliss?). ISME J 5:777–779

Gilbert JA, Jansson JK, Knight R (2014) The earth microbiome project: successes and aspirations. BMC Biol 12(1):69. https://doi.org/10.1186/s12915-014-0069-1

Glass EM, Wilkening J, Wilke A, Antonopoulos D, Meyer F (2010) Using the metagenomics RAST server (MG-RAST) for analyzing shotgun metagenomes. Cold Spring Harb Protoc. https://doi.org/10.1101/pdb.prot5368

Glick BR (2012) Plant growth-promoting bacteria: mechanisms and applications. Scientifica 2012:963401, 15 pp. https://doi.org/10.6064/2012/963401

Glick BR (2014) Bacteria with ACC deaminase can promote plant growth and help to feed the world. Microbiol Res 169:30–39

Gole J et al (2013) Massively parallel polymerase cloning and genome sequencing of single cells using nanoliter microwells. Nat Biotechnol 31:1126–1132

Goll J, Rusch DB, Tanenbaum DM, Thiagarajan M, Li K, Methé BA, Yooseph S (2010) METAREP: JCVI metagenomics reports-an open source tool for high-performance comparative metagenomics. Bioinformatics 26:2631–2632

Gori F, Folino G, Jetten MS, Marchiori E (2011) MTR: taxonomic annotation of short metagenomic reads using clustering at multiple taxonomic ranks. Bioinformatics 27:196–203

Gosalbes MJ, Durbán A, Pignatelli M, Abellan JJ, Jiménez-Hernández N, Pérez-Cobas AE, Moya A (2011) Metatranscriptomic approach to analyze the functional human gut microbiota. PLoS One 6:e17447. https://doi.org/10.1371/journal.pone.0017447

Grabherr MG, Haas BJ, Yassour M, Levin JZ, Thompson DA, Amit I, Chen Z (2011) Full-length transcriptome assembly from RNA-Seq data without a reference genome. Nat Biotechnol 29:644–652

Graham-Rowe D (2011) Beyond food versus fuel. Nature 474:S6–S8

Gruber-Dorninger C, Pester M, Kitzinger K, Savio DF, Loy A, Rattei T, Daims H (2015) Functionally relevant diversity of closely related *Nitrospira* in activated sludge. ISME J 9:643–655

Guangyin Z, Xueqin Lu, Takuro K, Gopalakrishnan K, Kaiqin X (2016) Promoted electromethanosynthesis in a two-chamber microbial electrolysis cells (MECs) containing a hybrid biocathode covered with graphite felt (GF). Chem Eng J 284:1146–1155

Guo J, Cole JR, Zhang Q, Brown CT, Tiedje JM (2016) Microbial community analysis with ribosomal gene fragments from shotgun metagenomes. Appl Environ Microbiol 82:157–166

Gygi SP, Corthals GL, Zhang Y, Rochon Y, Aebersol R (2000) Evaluation of two-dimensional gel electrophoresis-based proteome analysis technology. Proc Natl Acad Sci U S A 97:9390–9395

Haft DH, Selengut JD, White O (2003) The TIGRFAMs database of protein families. Nucleic Acids Res 1:371–373

Ham RG (1965) Clonal growth of mammalian cells in a chemically defined, synthetic medium. Proc Natl Acad Sci U S A 53:288–293

Handelsman J (2004) The ecologist's field guide to sequence-based identification of biodiversity. Microbiol Mol Biol Rev 68:669–685

Hanning I, Diaz-Sanchez S (2015) The functionality of the gastrointestinal microbiome in non-human animals. Microbiome 3:51

Harfouche A, Meilan R, Altman A (2011) Tree genetic engineering and applications to sustainable forestry and biomass production. Trends Biotechnol 29:9–17

Harrington ED, Arumugam M, Raes J, Bork P, Relman DA (2010) SmashCell: a software framework for the analysis of single-cell amplified genome sequences. Bioinformatics 26:2979–2980

Hayashi K (1991) PCR-SSCP: a simple and sensitive method for detection of mutations in the genomic DNA. PCR Methods Appl 1:34–38

He Z, Piceno Y, Deng Y, Xu M, Lu Z, DeSantis T, Andersen G, Sarah E, Hobbie SE, Reich PB, Zhou J (2012) The phylogenetic composition and structure of soil microbial communities shifts in response to elevated carbon dioxide. ISME J 6:259–272

Henson J, Tischler G, Ning Z (2012) Next-generation sequencing and large genome assemblies. Pharmacogenomics 13:901–915

Hernández-León R, Rojas-Solís D, Contreras-Pérez M, Orozco-Mosqueda MC, Macías-Rodríguez LI, Reyes-de la Cruz H, Valencia-Cantero E, Santoyo G (2015) Characterization of the antifungal and plant growth-promoting effects of diffusible and volatile organic compounds produced by Pseudomonas fluorescensstrains. Biol Control 81:83–92

Hettich RL, Chourey K, Jansson J, VerBerkmoes N, Shah M, Chavarria KL, Tom LM, Brodie EL (2010) Direct cellular Lysis/protein extraction protocol for soil metaproteomics. J Proteome Res 9:6615–6622

Hettich RL, Pan C, Chourey K, Giannone RJ (2013) Metaproteomics: harnessing the power of high performance mass spectrometry to identify the suite of proteins that control metabolic activities in microbial communities. Anal Chem 85:4203–4214

Hewson I, Fuhrman JA (2004) Richness and diversity of bacterioplankton species along an estuarine gradient in Moreton Bay, Australia. Appl Environ Microbiol 70:3425–3433

Higashi S, Barreto AMS, Cantão ME, Vasconcelos ATR (2012) Analysis of composition-based metagenomic classification. BMC Genomics 13:S1

Higuchi R, Dollinger G, Walsh PS, Griffith R (1992) Simultaneous amplification and detection of specific DNA sequences. Nat Biotechnol 10:413–417

Hoff KJ, Lingner T, Meinicke P, Tech M (2009) Orphelia: predicting genes in metagenomic sequencing reads. Nucleic Acids Res 37:W101–W105

Hooper SD, Dalevi D, Pati A, Mavromatis K, Ivanova NN, Kyrpides NC (2010) Estimating DNA coverage and abundance in metagenomes using a gamma approximation. Bioinformatics 26:295–301

Horton M, Bodenhausen N, Bergelson J (2010) MARTA: a suite of Java-based tools for assigning taxonomic status to DNA sequences. Bioinformatics 26:568–569

Hou Y, Song L, Zhu P, Zhang B, Tao Y, Xu X, Wu H (2012) Single-cell exome sequencing and monoclonal evolution of a JAK2-negative myeloproliferative neoplasm. Cell 148(5):873–885

Hou J, Liu W, Wang B, Wang Q, Luo Y, Franks AE (2015) PGPR enhanced phytoremediation of petroleum contaminated soil and rhizosphere microbial community response. Chemosphere 138:592–598

Huang XD, El-Alawi Y, Penrose DM, Glick BR, Greenberg BM (2004) Responses of three grass species to creosote during phytoremediation. Environ Pollut 130:453–463

Huang CJ, Tsay JF, Chang SY, Yang HP, Wu WS, Chen CY (2012) Dimethyl disulfide is an induced systemic resistance elicitor produced by Bacillus cereus C1L. Pest Manag Sci 68:1306–1310

Huang XF, Chaparro JM, Reardon KF, Zhang R, Shen Q, Vivanco JM (2014) Rhizosphere interactions: root exudates, microbes, and microbial communities. Botany 92:267–275

Hunter S, Corbett M, Denise H, Fraser M, Gonzalez-Beltran A, Hunter C, Jones P, Leinonen R, McAnulla C, Maguire E, Maslen J, Mitchell A, Nuka G, Oisel A, Pesseat S, Radhakrishnan R, Rocca-Serra P, Scheremetjew M, Sterk P, Vaughan D, Cochrane G, Field D, Sansone SA (2014a) metagenomics-a new resource for the analysis and archiving of metagenomic data. Nucleic Acids Res 42:D600–D606

Hunter S, Corbett M, Denise H, Fraser M, Gonzalez-Beltran A, Hunter C, Jones P, Leinonen R, McAnulla C, Maguire E, Maslen J, Mitchell A, Nuka G, Oisel A, Pesseat S, Radhakrishnan R, Rocca-Serra P, Scheremetjew M, Sterk P, Vaughan D, Cochrane G, Field D, Sansone SA (2014b) EBI metagenomics-a new resource for the analysis and archiving of metagenomic data. Nucleic Acids Res 42:D600–D606

Huse SM, Mark Welch DB, Voorhis A, Shipunova A, Morrison HG, Eren AM, Sogin ML (2014) VAMPS: a website for visualization and analysis of microbial population structures. BMC Bioinform 15:41

Huson DH, Auch AF, Qi J, Schuster SC (2007) MEGAN analysis of metagenomic data. Genome Res 17:377–386

Hussa EA, Goodrich-Blair H (2013) It takes a village: ecological and fitness impacts of multipartite mutualism. Annu Rev Microbiol 67:161–178

Huyghe A, Francois P, Schrenzel J (2009) Characterization of microbial pathogens by DNA microarrays. Infect Genet Evol 9:987–995

Hyland C, Pinney JW, McConkey GA, Westhead DR (2006) metaSHARK: a WWW plateform for interactive exploration of metabolic networks. Nucleic Acids Res 34:W725–W728

Ishii K, Mußmann M, MacGregor BJ, Amann R (2004) An improved fluorescence in situ hybridization protocol for the identification of bacteria and archaea in marine sediments. FEMS Microbiol Ecol 50:203–212

Iwasaki Y, Abe T, Wada K, Wada Y, Ikemura T (2013) A novel bioinformatics strategy to analyze microbial big sequence data for efficient knowledge discovery: batch-learning self-organizing map (BLSOM). Microorganisms 1:137–157

Janczyk P, Pieper R, Smidt H, Wolfgang B (2010) Souffrant effect of alginate and inulin on intestinal microbial ecology of weanling pigs reared under different husbandry conditions. FEMS Microbiol Ecol 72:132–142

Jansson J (2011) Soil microbes: metagenomic approaches. https://eesa.lbl.gov/soil-microbes-metagenomic-approaches/2011

Jansson JK (2013) The life beneath our feet. Nature 494:40

Ji SH, Gururani MA, Chun SC (2014) Isolation and characterization of plant growth promoting endophytic diazotrophic bacteria from Korean rice cultivars. Microbiol Res 169:83–98

Jiang L, He L, Fountoulakis M (2004) Comparison of protein precipitation methods for sample preparation prior to proteomic analysis. J Chromatogr A 1023:317–320

Jousset A, Lara E, Nikolausz M, Harms H, Chatzinotas A (2010) Application of the denaturing gradient gel electrophoresis (DGGE) technique as an efficient diagnostic tool for ciliate communities in soil. Sci Total Environ 408:1221–1225

Kakar KU, Duan Y-P, Nawaz Z, Sun G, Almoneafy AA, Hassan MA, Elshakh A, Li B, Xie G-L (2014) A novel rhizobacterium Bk7 for biological control of brown sheath rot of rice caused by Pseudomonas fuscovaginaeand its mode of action. Eur J Plant Pathol 138:819–834

Kalisky T, Blainey P, Quake SR (2011) Genomic analysis at the single-cell level. Annu Rev Genet 45:431–445

Kamke J, Bayer K, Woyke T, Hentschel U (2012) Exploring symbioses by single-cell genomics. Biol Bull 223:30–43

Kan J, Hanson TE, Ginter JM, Wang K, Chen F (2005) Metaproteomic analysis of Chesapeake Bay microbial communities. Saline Syst 1:7–10

Kanehisa M, Goto S (2000) KEGG: kyoto encyclopedia of genes and genomes. Nucleic Acids Res 28:27–30

Kaur A, Chaudhary A, Kaur A, Choudhary R, Kaushik R (2005) Phospholipid fatty acid – a bioindicator of environment monitoring assessment in soil ecosystem. Curr Sci 89:1103–1112

Keegan KP, Glass EM, Meyer F (2016) MG-RAST, a metagenomics service for analysis of microbial community structure and function. Methods Mol Biol 1399:207–233

Keiblinger KM, Wilhartitz IC, Schneider T, Roschitzki B, Schmid E, Eberl L, Riedel K, Zechmeister-Boltenstern S (2012) Soil metaproteomicse comparative evaluation of protein extraction protocols. Soil Biol Biochem 54:14–24

Kembel SW, Eisen JA, Pollard KS, Green JL (2011) The phylogenetic diversity of metagenomes. PLoS One 6:e23214

Khaksar G, Treesubsuntorn C, Thiravetyan P (2016) Effect of endophytic Bacillus cereus ERBP inoculation into non-native host: potentials and challenges for airborne formaldehyde removal. Plant Physiol Biochem 107:326–336

Khan Z, Roman D, Kintz T, delas Alas M, Yap R, Doty S (2014) Degradation, phytoprotection and phytoremediation of phenanthrene by endophyte Pseudomonas putida, PD1. Environ Sci Technol 48:12221–12228

Khindaria A, Grover TA, Aust SD (1995) Reductive dehalogenation of aliphatic halocarbons by lignin peroxidase of Phanerochaete chrysosporium. Environ Sci Technol 29:719–725

Kim HJ, Choi HS, Yang SY, Kim IS, Yamaguchi T, Sohng JK, Park SK, Kim JC, Lee CH, Gardener BM et al (2014) Both extracellular chitinase and a new cyclic lipopeptide, chromobactomycin, contribute to the biocontrol activity of *Chromobacterium* sp. C61. Mol Plant Pathol 15:122–132

Kind J, Pagie L, Ortabozkoyun H, Boyle S, de Vries SS et al (2013) Single-cell dynamics of genome-nuclear lamina interactions. Cell 153:178–192

Klose J (1975) Protein mapping by combined isoelectric focusing and electrophoresis of mouse tissues. A novel approach to testing for induced point mutations in mammals. Humangenetik 26:231–243

Knight R, Vrbanac A, Taylor BC, Aksenov A, Callewaert C, Debelius J, Melnik AV (2018) Best practices for analysing microbiomes. Nat Rev Microbiol 16:410–422

Knights D, Kuczynski J, Charlson ES, Zaneveld J, Mozer MC, Collman RG, Bushman FD, Knight R, Kelley ST (2011) Bayesian community-wide culture-independent microbial source tracking. Nat Methods 8:761–763

Ko CH, Yu FC, Chang FC, Yang BY, Chen WH, Hwang WS, Tu TC (2017) Bioethanol production from recovered napier grass with heavy metals. J Environ Manag 203:1005–1010

Köcher T, Pichler P, Swart R, Mechtler K (2012) Analysis of protein mixtures from whole-cell extracts by single-run nanoLC-MS/MS using ultralong gradients. Nat Protoc 7:882–890

Kohler J, Caravaca F, Azcon R, Diaz G, Roldan A (2016) Suitability of the microbial community composition and function in a semiarid mine soil for assessing phytomanagement practices based on mycorrhizal inoculation and amendment addition. J Environ Manag 169:236–246

Kolmeder CA, de Been M, Nikkilä J, Ritamo I, Mättö J, Valmu L, Salojärvi J, Palva A, Salonen A, de Vos WM (2012) Comparative metaproteomics and diversity analysis of human intestinal microbiota testifies for its temporal stability and expression of core functions. PLoS One 7:e29913–e29910

Korf BR, Rehm HL (2013) New approaches to molecular diagnosis. JAMA 309:1511–1521

Kovacs A, Yacoby K, Gophna U (2010a) A systematic assessment of automated ribosomal intergenic spacer analysis (ARISA) as a tool for estimating bacterial richness. Res. Microbiology 161:192–197

Kovacs A et al (2010b) Genotype is a stronger determinant than sex of the mouse gut microbiota. Microbial Ecol 1:6

Krey T, Vassilev N, Baum C, Eichler-Löbermann B (2013) Effects of long-term phosphorus application and plant-growth promoting rhizobacteria on maize phosphorus nutrition under field conditions. Eur J Soil Biol 55:124–130

Kristiansson E, Hugenholtz P, Dalevi D (2009) ShotgunFunctionalizeR: an R-package for functional comparison of metagenomes. Bioinformatics 25:2737–2738

Kultima JR, Sunagawa S, Li J, Chen W, Chen H, Mende DR, Arumugam M, Pan Q, Liu B, Qin J, Wang J, Bork P (2012) MOCAT: a metagenomics assembly and gene prediction toolkit. PLoS One 7:e47656

Kumar KV, Singh N, Behl HM, Srivastava S (2008) Influence of plant growth promoting bacteria and its mutant on heavy metal toxicity in *Brassica juncea* grown in fly ash amended soil. Chemosphere 72:678–683

Kunin V, Copeland A, Lapidus A, Mavromatis K, Hugenholtz P (2008) A bioinformatician's guide to metagenomics. Microbiol Mol Biol Rev 72:557–578

Lacerda CMR, Choe LH, Reardon KF (2007) Metaproteomic analysis of a bacterial community response to cadmium exposure. J Proteome Res 6:1145–1152

Landry ZC, Giovanonni SJ, Quake SR, Blainey PC (2013) Optofluidic cell selection from complex microbial communities for single-genome analysis. Methods Enzymol 531:61–90

Langmead B, Salzberg SL (2012) Fast gapped-read alignment with Bowtie2. Nat Methods 9:357–359

Lapidot D, Dror R, Vered E, Mishli O, Levy D, Helman Y (2014) Disease protection and growth promotion of potatoes (*Solanum tuberosum* L.) by *Paenibacillus dendritiformis*. Plant Pathol. https://doi.org/10.1111/ppa.1228

Larsen PE, Collart FR, Field D, Meyer F, Keegan KP, Henry CS, McGrath J, Quinn J, Gilbert JA (2011) Predicted Relative Metabolomic Turnover (PRMT): determining metabolic turnover from a coastal marine metagenomic dataset. Microb Inf Exp 1:4

Laserson J, Jojic V, Koller D (2011) Genovo: de novo assembly for metagenomes. J Comput Biol 18:429–443

Lasken RS (2012) Genomic sequencing of uncultured microorganisms from single cells. Nat Rev Microbiol 10:631–640. https://doi.org/10.1038/nrmicro2857

Leary DH, Hervey WJ, Li RW, Deschamps JR, Kusterbeck AW, Vora GJ (2012) Method development for metaproteomic analyses of marine biofilms. Anal Chem 84:4006–4013

Lecault V, White AK, Singhal A, Hansen CL (2012) Microfluidic single cell analysis: from promise to practice. Curr Opin Chem Biol 16:381–390

Lee KH (2001) Proteomics: a technology-driven and technology-limited discovery science. Trends Biotechnol 19:217–222

Lee DH, Zo YG, Kim SJ (1996) Nonradioactive method to study genetic profiles of natural bacterial communities by PCR–single-strand conformation polymorphism. Appl Environ Microbiol 62:3112–3120

Lee CK, Barbier BK, Bottos EM, McDonald IR, Cary SC (2012) The Inter-Valley Soil Comparative Survey: the ecology of Dry Valley edaphic microbial communities. ISME J 6:1046–1057

Lehman RM, Cambardella CA, Stott DE, Acosta-Martinez V, Manter DK, Buyer JS, Maul JE, Smith JL, Collins HP, Halvorson JP (2015) Understanding and enhancing soil biological health: the solution for reversing soil degradation. Sustainability 7:988–1027

Lehninger AL (1965) Bioenergetics: The molecular basis of biological energy transformations. W. A. Benjamin, New York. pp xv, 258

Lester ED, Satomi M, Ponce A (2007) Microflora of extreme arid Atacama Desert soils. Soil Biol Biochem 39:704–708

Leung K et al (2012) A programmable droplet-based microfluidic device applied to multiparameter analysis of single microbes and microbial communities. Proc Natl Acad Sci U S A 109:7665–7670

Leung ML, Wang Y, Waters J, Navin NES (2015) single nucleus exome sequencing. Genome Biol 16:55

Li W (2009) Analysis and comparison of very large metagenomes with fast clustering and functional annotation. BMC Bioinform 10:359. https://doi.org/10.1186/1471-2105-10-359

Li R, Zhu H, Ruan J, Qian W, Fang X, Shi Z, Li Y, Li S, Shan G, Kristiansen K, Li S, Yang H, Wang J, Wang J (2010) De novo assembly of human genomes with massively parallel short read sequencing. Genome Res 20:265–272

Li H-Y et al (2012) Endophytes and their role in phytoremediation. Fungal Divers 54:11–18

Li X, Sun J, Wang H, Li X, Wang J, Zhang H (2017) Changes in the soil microbial phospholipid fatty acid profile with depth in three soil types of paddy fields in China. Geoderma 290:69–74

Lichter P, Ledbetter SA, Ledbetter DH, Ward DC (1990) Fluorescence in situ hybridization with Alu and L1 polymerase chain reaction probes for rapid characterization of human chromosomes in hybrid cell lines. Proc Natl Acad Sci U S A 87:6634–6638

Likar M, Stres B, Rusjan D, Potisek M, Regvar M (2017) Ecological and conventional viticulture gives rise to distinct fungal and bacterial microbial communities in vineyard soils. Appl Soil Ecol 113:86–95

Lin HH, Liao YC (2016) Accurate binning of metagenomic contigs via automated clustering sequences using information of genomic signatures and marker genes. Sci Rep 6:24175

Liu B, Gibbons T, Ghodsi M, Treangen T, Pop M (2011) Accurate and fast estimation of taxonomic profiles from metagenomic shotgun sequences. BMC Genomics 2:S4

Liu W, Sun J, Ding L, Luo Y, Chen M, Tang C (2013) Rhizobacteria (*Pseudomonas sp. SB*) assist phytoremediation of oily-sludge-contaminated soil by tall fescue (*Testuca arundinacea* L.). Plant Soil 371:533–542

Liu Y, Wang P, Crowley D, Liu X, Chen J, Li L, Zheng J, Zhang X, Zheng J, Pan G (2016a) Methanogenic abundance and changes in community structure along a rice soil chronose-quence from east China. Eur J Soil Sci. https://doi.org/10.1111/ejss.12348

Liu Y, Yang D, Zhang N, Chen L, Cui Z, Shen Q, Zhang R (2016b) Characterization of uncultured genome fragment from soil metagenomic library exposed rare mismatch of internal tetranucle-otide frequency. Front Microbiol 7:2081

Lovett M (2013) The applications of single-cell genomics. Hum Mol Genet 22:R22–R26

Lucas JA, García-Cristobal J, Bonilla A, Ramos B, Gutierrez-Mañero J (2014) Beneficial rhizo-bacteria from rice rhizosphere confers high protection against biotic and abiotic stress inducing systemic resistance in rice seedlings. Plant Physiol Biochem 82:44–53

Lukow T, Dunceld PF, Liesack W (2000) Use of the T-RFLP technique to assess spatial and tem-poral changes in the bacterial community structure within an agricultural soil planted with transgenic and non-transgenic potato plants. FEMS Microbiol Ecol 32:241–247

Luo C, Zhou H, Wang X, Zhang R, Xiang Y (2014) Bacillomycin L and surfactin contribute syn-ergistically to the phenotypic features of Bacillus subtilis 916 and the biocontrol of rice sheath blight induced by Rhizoctonia solani. Appl Microbiol Biotechnol. https://doi.org/10.1007/s00253-014-6195-4

Lupwayi NZ, Lamey FJ, Blackshaw RE, Kanashiro DA, Pearson DC (2017) Phospholipid fatty acid biomarkers show positive soil microbial community responses to conservation soil man-agement of irrigated crop rotations. Soil Tillage Res 168:1–10

Ma Y, Rajkumar M, Freitas H (2009) Isolation and characterization of Ni mobilizing PGPB from serpentine soils and their potential in promoting plant growth and Ni accumulation by Brassica spp. Chemosphere 75:719–725

Ma Y, Rajkumar M, Zhang C, Freitas H (2016) Inoculation of Brassica oxyrrhina with plant growth promoting bacteria for the improvement of heavy metal phytoremediation under drought con-ditions. J Hazard Mater 320:36–44

Macaulay IC, Voet T (2014) Single cell genomics: advances and future perspectives. PLoS Genet 10:e1004126

Macosko EZ et al (2015) Highly parallel genome-wide expression profiling of individual cells using nanoliter droplets. Cell 161:1202–1214

Maldonado-Mendoza IE, Galindo-flores H, Lopez-meyer M (2009) An introduction to metagenom-ics. In: Chauhan AK, Varma A (eds) A textbook of molecular biotechnology. I. K. International Publishing House Pvt Ltd, New Delhi

Malghani S, Reim A, Von Fischer J, Conrad R, Kuebler K, Trumbore SE (2016) Soil methanotroph abundance and community composition are not influenced by substrate availability in labora-tory incubations. Soil Biol Biochem 101:184–194

Manefield M, Whiteley AS, Griffiths RI, Bailey MJ (2002) RNA stable isotope probing, a novel means of linking microbial community function to phylogeny. Appl Environ Microbiol 68:5367–5373

Mann M, Pandey A (2001) Use of mass spectrometry-derived data to annotate nucleotide and protein sequence databases. Trends Biochem Sci 26:54–61. https://doi.org/10.1016/S0968-0004(00)01726-6

Marco-Sola S, Sammeth M, Guigó R, Ribeca P (2012) The GEM mapper: fast, accurate and versatile alignment by filtration. Nat Methods 9:1185–1188

Marcy Y, Ishoey T, Lasken RS, Stockwell TB, Walenz BP, Halpern AL, Beeson KY, Goldberg SMD, Quake SR (2007) Nanoliter reactors improve multiple displacement amplification of genomes from single cells. PLoS Genet 3:1702–1708

Markowicz A, Cycoń M, Piotrowska-Seget Z (2016) Microbial community structure and diversity in long-term hydrocarbon and heavy metal contaminated soils. Int J Environ Res 10:321–332

Markowitz VM, Chen IM, Chu K, Szeto E, Palaniappan K, Grechkin Y, Ratner A, Jacob B, Pati A, Huntemann M, Liolios K, Pagani I, Anderson I, Mavromatis K, Ivanova NN, Kyrpides

NC (2012) IMG/M: the integrated metagenome data management and comparative analysis system. Nucleic Acids Res 40:D123–D129

Maron PA, Ranjard L, Mougel C, Lemanceau P (2007) Metaproteomics: a new approach for studying functional microbial ecology. Microb Ecol 53:486–493

Martin HG, Ivanova N, Kunin V, Warnecke F, Barry KW, McHardy AC, Yeates C, He SM, Salamov AA, Szeto E, Dalin E, Putnam NH, Shapiro HJ, Pangilinan JL, Rigoutsos I, Kyrpides NC, Blackall LL, McMahon KD, Hugenholtz P (2006) Metagenomic analysis of two enhanced biological phosphorus removal (EBPR) sludge communities. Nat Biotechnol 24:1263–1269

Martínez-García M, Santos F, Moreno-Paz M, Parro V, Antón J (2014) Unveiling viral-host interactions within the 'microbial dark matter'. Nat Commun 5:4542

Martin-Laurent F, Philippot L, Hallet S, Chaussod R, Germon JC, Soulas G, Catroux G (2001) DNA extraction from soils: old bias for new microbial diversity analysis methods. Appl Environ Microbiol 67:2354–2359

Mathé C, Sagot MF, Schiex T, Rouzé P (2002) Survey and summary: current methods of gene prediction, their strengths and weaknesses. Nucleic Acids Res 30:4103–4117

McElroy KE, Luciani F, Thomas T (2012) GemSIM: general, error-model based simulator of next-generation sequencing data. BMC Genomics 13:74

McGenity TJ, Crombie AT, Murrell JC (2018) Microbial cycling of isoprene, the most abundantly produced biological volatile organic compound on Earth. ISME J 12:931–941

McGrath KC, Mondav R, Sintrajaya R, Slattery B, Schmidt S, Schenk PM (2010) Development of an environmental functional gene microarray for soil microbial communities. Appl Environ Microbiol 76:7161–7170

McHardy AC, Martin HG, Tsirigos A, Hugenholtz P, Rigoutsos I (2007) Accurate phylogenetic classification of variable-length DNA fragments. Nat Methods 4(63–72):10

McKenney DW, Yemshanov D, Fraleigh S, Allen D, Preto F (2011) An economic assessment of the use of short-rotation coppice woody biomass to heat greenhouses in southern Canada. Biomass Bioenergy 35:374–384

Meers E, Van Slyckena S, Adriaensenb K, Ruttensb A, Vangronsveld J, Du Laing G, Witters N, Thewysb T, Tack FM (2010) The use of bio-energy crops (*Zea mays*) for 'phytoattenuation' of heavy metals on moderately contaminated soils: a field experiment. Chemosphere 78:35–41

Melcher U, Verma R, Schneider WL (2014) Metagenomic search strategies for interactions among plants and multiple microbes. Front Plant Sci 5:268

Mendoza MLZ, Sicheritz-Pontén T, Gilbert MTP (2015) Environmental genes and genomes: understanding the differences and challenges in the approaches and software for their analyses. Brief Bioinform 16:745–758

Meyer F, Paarmann D, D'Souza M, Olson R, Glass EM, Kubal M, Paczian T, Rodriguez A, Stevens R, Wilke A, Wilkening J, Edwards RA (2008) The metagenomics RAST server – a public resource for the automatic phylogenetic and functional analysis of metagenomes. BMC Bioinform 9:386

Mijangos I, Becerril JM, Albizu I, Epelde L, Garbisu C (2009) Effects of glyphosate on rhizosphere soil microbial communities under two different plant compositions by cultivation-dependent and independent methodologies. Soil Biol Biochem 41:505–513

Miller JR, Koren S, Sutton G (2010) Assembly algorithms for next-generation sequencing data. Genomics 95:315–327

Mocali S, Benedetti A (2010) Exploring research frontiers in microbiology: the challenge of metagenomics in soil microbiology. Res Microbiol 161:497–505

Mohammed MH, Ghosh TS, Reddy RM, Reddy CV, Singh NK, Mande SS (2011) INDUS – a composition-based approach for rapid and accurate taxonomic classification of metagenomic sequences. BMC Genomics 3:S4

Molina LG, Cordenonsi da Fonseca G, Morais GLD, de Oliveira LFV, Carvalho JBD, Kulcheski FR, Margis R (2012) Metatranscriptomic analysis of small RNAs present in soybean deep sequencing libraries. Genet Mol Biol 35:292–303

Monzoorul Haque M, Ghosh TS, Komanduri D, Mande SS (2009) SOrt-ITEMS: Sequence orthology based approach for improved taxonomic estimation of metagenomic sequences. Bioinformatics 25:1722–1730

Morales SE, Cosart T, Holben WE (2010) Bacterial gene abundances as indicators of greenhouse gas emission in soils. ISME J 4:799–808

Moran MA (2009) Metatranscriptomics: eavesdropping on complex microbial communities. Microbe 4:329–335

Morgan XC, Huttenhower C (2014) Meta-omic analytic techniques for studying the intestinal microbiome. Gastroenterology 146:1437–1448

Mossa AW, Dickinson MJ, West HM, Young SD, Crout NM (2017) The response of soil microbial diversity and abundance to long-term application of biosolids. Environ Pollut 224:16–25

Mukherjee S, Stamatis D, Bertsch J, Ovchinnikova G, Verezemska O, Isbandi M, Thomas AD, Ali R, Sharma K, Kyrpides NC, Reddy TB (2017) Data updates and feature enhancements. Nucleic Acids Res 45:D446–D456

Muller J, Szklarczyk D, Julien P, Letunic I, Roth A, Kuhn M, Powell S, von Mering C, Doerks T, Jensen LJ, Bork P (2010) extending the evolutionary genealogy of genes with enhanced non-supervised orthologous groups, species and functional annotations. Nucleic Acids Res 38:D190–D195

Muyzer G (1999) DGGE/TGGE a method for identifying genes from natural ecosystems. Curr Opin Microbiol 2:317–322

Nadeem SM, Ahmad M, Zahir ZA, Javaid A, Ashra M (2014) The role of mycorrhizae and plant growth promoting rhizobacteria (PGPR) in improving crop productivity under stressful environments. Biotechnol Adv 32:429–448

Nalbantoglu OU, Way SF, Hinrichs SH, Sayood K (2011) RAIphy: phylogenetic classification of metagenomics samples using iterative refinement of relative abundance index profiles. BMC Bioinform 12:41. https://doi.org/10.1186/1471-2105-12-41

Namiki T, Hachiya T, Tanaka H, Sakakibara Y (2012) MetaVelvet: an extension of Velvet assembler to de novo metagenome assembly from short sequence reads. Nucleic Acids Res 40:e155

Nannipieri P, Smalla K (2006) Role of stabilised enzymes in microbial ecology and enzyme extraction from soil with potential applications in soil proteomics nucleic acids and proteins in soil. Springer, Berlin/Heidelberg, pp 75–94

Nautiyal CS, Srivastava S, Chauhan PS, Seem K, Mishra A, Sopory SK (2013) Plant growth-promoting bacteria Bacillus amyloliquefaciens NBRISN13 modulates gene expression profile of leaf and rhizosphere community in rice during salt stress. Plant Physiol Biochem 66:1–9

Navin N et al (2011) Tumour evolution inferred by single-cell sequencing. Nature 472:90–94

Neiverth A, Delai S, Garcia DM, Saatkamp K, Souza EM, Pedrosa FO, Guimarães VF, Santos MF, Vendruscolo ECG, Costa ACT (2014) Performance of different wheat genotypes inoculated with the plant growth promoting bacterium Herbaspirillum seropedicae. Eur J Soil Biol 64:1–5. https://doi.org/10.1016/j.ejsobi.2014.07.001

Neufeld JD, Dumont MG, Vohra J, Murrell JC (2007) Methodological considerations for the use of stable isotope probing in microbial ecology. Microb Ecol 53:435–442

Noguchi H, Taniguchi T, Itoh T (2008) MetaGeneAnnotator: detecting species-specific patterns of ribosomal binding site for precise gene prediction in anonymous prokaryotic and phage genomes. DNA Res 15:387–396

Novotný Č, Vyas BRM, Erbanova P, Kubatova A, Šašek V (1997) Removal of PCBs by various white rot fungi in liquid cultures. Folia Microbiol 42:136–140

NRSC I (2011) Wastelands Atlas of India 2011: change analysis based on temporal satellite data of 2005–06 and 2008–09. National Remote Sensing Centre (NRSC), Hyderabad

Nunan N, Daniell TJ, Singh BK, Papert A, McNicol JW, Prosser JI (2005) Links between plant and rhizoplane bacterial communities in grassland soils, characterized using molecular techniques. Appl Environ Microbiol 71:6784–6792

O'Farrell PH (1975) High resolution two-dimensional electrophoresis of proteins. J Biol Chem 250:4007–4021

Ochsenreiter T, Selezi D, Quaiser A, Bonch-Osmolovskaya L, Schleper C (2003) Diversity and abundance of Crenarchaeota in terrestrial habitats studied by 16S RNA surveys and real time PCR. Environ Microbiol 5:787–797

Ogier JC, Son O, Gruss A, Tailliez P, Delacroix-Buchet A (2002) Identification of the bacterial microflora in dairy products by temporal temperature gradient gel electrophoresis. Appl Environ Microbiol 68:3691–3701

Ogunseitan OA (1993) Direct extraction of proteins from environmental samples. J Microbiol Methods 17:273–281

Ogunseitan OA (1996) Protein profile in cultivated and native freshwater microorganisms exposed to chemical environmental pollutants. Microb Ecol 31:291–304

Ogunseitan OA (1997) Direct extraction of catalytic proteins from natural microbial communities. J Microbiol Methods 28:55–63

Ogunseitan O (2005) Microbial diversity: form and function in prokaryotes. Blackwell Science Ltd, Malden, p 142

Overbeek MV, Kusuma WA, Buono A (2013) Clustering metagenome fragments using growing self organizing map. In: 2013 International conference on advanced computer science and information systems (ICACSIS). IEEE, pp 285–289

Pandey A, Lewitter F (1999) Nucleotide sequence databases: a gold mine for biologists. Trends Biochem Sci 24:276–280

Pandey A, Mann M (2000) Proteomics to study genes and genomes. Nature 405:837–846

Parks DH, Tyson GW, Hugenholtz P, Beiko RG (2014) STAMP: statistical analysis of taxonomic and functional profiles. Bioinformatics 30:3123–3124

Pati A, Heath LS, Kyrpides NC, Ivanova N (2011) ClaMS: a classifier for metagenomic sequences. Stand Genomic Sci 5:248–253

Patil KR, Roune L, McHardy AC (2012) The PhyloPythiaS web server for taxonomic assignment of metagenome sequences. PLoS One 7(6):e38581. https://doi.org/10.1371/journal.pone.0038581

Paulson JN, Pop M, Bravo HC (2011) Metastats: an improved statistical method for analysis of metagenomic data. Genome Biol 12:P17

Peano C, Pietrelli A, Consolandi C, Rossi E, Petiti L, Tagliabue L, Landini P (2013) An efficient rRNA removal method for RNA sequencing in GC-rich bacteria. Microb Inform Exp 3:1. https://doi.org/10.1186/2042-5783-3-1

Pedersen S, Bloch PL, Reeh S, Neidhardt FC (1978) Patterns of protein synthesis in *E. coli*: a catalog of the amount of 140 individual proteins at different growth rates. Cell 14:179–190

Pell J, Hintze A, Canino-Koning R, Howe A, Tiedje JM, Brown CT (2012) Scaling metagenome sequence assembly with probabilistic de Bruijn graphs. Proc Natl Acad Sci U S A 109:13272–13277

Peng J, Elias JE, Thoreen CC, Licklider LJ, Gygi SP (2003) Evaluation of multidimensional chromatography coupled with tandem mass spectrometry (LC/LC−MS/MS) for large-scale protein analysis: the yeast proteome. J Proteome Res 2:43–45

Peng Y, Henry CM, Leung S, Yiu M, Francis Y, Chin L (2011) Meta-IDBA: a de Novo assembler for metagenomic data. Bioinformatics 27:i94–i101

Peng Y, Leung HC, Yiu SM, Chin FY (2012) IDBA-UD: a de novo assembler for single-cell and metagenomic sequencing data with highly uneven depth. Bioinformatics 28:1420–1428

Perez-Cobas AE, Gosalbes MJ, Friedrichs A, Knecht H, Artacho A, Eismann K, Otto W, Rojo D, Bargiela R, von Bergen M, Neulinger SC, Däumer C, Heinsen FA, Latorre A, Barbas C, Seifert J, dos Santos VM, Ott SJ, Ferrer M, Moya A (2013) Gut microbiota disturbance during antibiotic therapy: a multi-omic approach. Gut 62:1591–1601

Pérez-Montaño F, Alías-Villegas C, Bellogín RA, del Cerro P, Espuny MR, Jiménez-Guerrero I, López-Baena FJ, Ollero FJ, Cubo T (2014) Plant growth promotion in cereal and leguminous agricultural important plants: from microorganism capacities to crop production. Microbiol Res 169:325–336

Perkel JM (2012) Single-cell genomics: defining microbiology's dark matter. BioTechniques 52:301–303

Pernthaler A, Pernthaler J, Amann R (2002) Fluorescence in situ hybridization and catalyzed reporter deposition for the identification of marine bacteria. Appl Environ Microbiol 68:3094–3101

Philippot L, Raaijmakers JM, Lemanceau P, van der Putten WH (2013) Going back to the roots: the microbial ecology of the rhizosphere. Nat Rev Microbiol 11:789–799

Pointing SB, Belnap J (2012) Microbial colonization and controls in dryland systems. Nat Rev Microbiol 10:551–562

Pointing SB, Chana Y, Lacapa DC, Laua MCY, Jurgens JA, Farrell RL (2010) Highly specialized microbial diversity in hyper-arid polar desert. Proc Natl Acad Sci U S A 107:1254–1254

Ponomarova O, Patil KR (2015) Metabolic interactions in microbial communities: untangling the Gordian knot. Curr Opin Microbiol 27:37–44

Poretsky RS, Bano N, Buchan A, LeCleir G, Kleikemper J, Pickering M, Hollibaugh JT (2005) Analysis of microbial gene transcripts in environmental samples. Appl Environ Microbiol 71:4121–4126

Potshangbam M, Devi SI, Sahoo D, Strobel GA (2017) Functional characterization of endophytic fungal community associated with *Oryza sativa* L. and *Zea mays* L. Front Microbiol. https://doi.org/10.3389/fmicb.2017.00325

Power AG (2010) Ecosystem services and agriculture: tradeoffs and synergies. Philos Trans R Soc 365:2959–2971

Prakash O, Sharma R, Rahi P, Karthikeyan N (2014) Role of microorganisms in plant nutrition and health. In: Nutrient use efficiency: from basics to advances, pp 125–161. Springer-Nature, Switzerland

Pride DT, Schoenfeld T (2008) Genome signature analysis of thermal virus metagenomes reveals Archaea and thermophilic signatures. BMC Genomics 9:420

Priya H, Prasanna R, Ramakrishnan B, Bidyarani N, Babu S, Thapa S, Renuka N (2015) Influence of cyanobacterial inoculation on the culturable microbiome and growth of rice. Microbiol Res. https://doi.org/10.1016/j.micres.2014.12.011

Props R, Kerckhof FM, Rubbens P, De Vrieze J, Sanabria EH, Waegeman W, Monsieurs P, Hammes F, Boon N (2017) Absolute quantification of microbial taxon abundances. ISME J 11:584–587

Pruesse E, Quast C, Knittel K, Fuchs BM, Ludwig W, Peplies J, Glöckner FO (2007) SILVA: a comprehensive online resource for quality checked and aligned ribosomal RNA sequence data compatible with ARB. Nucleic Acids Res 35:7188–7196

Quemeneur M, Garrido F, Billard P, Breeze D, Leyyal C, Jauzeind M, Joulianb C (2016) Community structure and functional arrA gene diversity associated with arsenic reduction and release in an industrially contaminated soil. Geomicrobiol J 33:839–849

Ragauskas AJ, Williams CK, Davison BH, Britovsek G, Cairney J, Eckert CA et al (2006) The path forward for biofuels and biomaterials. Science 311:484e9

Rajendhran J, Gunasekaran P (2011) Microbial phylogeny and diversity: small subunit ribosomal RNA sequence analysis and beyond. Microbiol Res 166:99–110

Rajkumar M, Nagendran R, Lee KJ, Lee WH, Kim SZ (2006) Influence of plant growth promoting bacteria and Cr6+ on the growth of Indian mustard. Chemosphere 62:741–748

Ram RJ, VerBerkmoes NC, Thelen MP, Tyson GW, Baker BJ, Blake RC II, Shah M, Hettich RL, Banfield JF (2005) Community proteomics of a natural microbial biofilm. Science 308:1915–1920

Ramachandran N, Hainsworth E, Bhullar B, Eisenstein S, Rosen B, Lau AY, Walter JC, LaBaer J (2004) Self-assembling protein microarrays. Science 305:86–90

Ramesh A, Sharma SK, Sharma MP, Yadav N, Joshi OP (2014) Inoculation of zinc solubilizing *Bacillus aryabhattaistrains* for improved growth, mobilization and biofortification of zinc in soybean and wheat cultivated in Vertisols of Central India. Appl Soil Ecol 73:87–96

Rana A, Joshi M, Prasanna R, Singh Y, Nain L (2012) Biofortification of wheat through inoculation of plant growth promoting rhizobacteria and cyanobacteria. Eur J Soil Biol 50:118–126

Randle-Boggis RJ, Helgason T, Sapp M, Ashton PD (2016) Evaluating techniques for metagenome annotation using simulated sequence data. FEMS Microbiol Ecol 92:95

Rastogi G, Sani RK (2011) Molecular techniques to assess microbial community structure, function, and dynamics in the environment, pp 29–57. Springer-Nature, Switzerland

Ratzke C, Denk J, Gore J (2018) Ecological suicide in microbes. Nat Ecol Evol 2:867–872

Raynaud X, Nunan N (2014) Spatial ecology of bacteria at the microscale in soil. PLoS One 9:e87217

Rho M, Tang H, Ye Y (2010) FragGeneScan: predicting genes in short and error-prone reads. Nucleic Acids Res 38:e191

Richter DC, Ott F, Auch AF, Schmid R, Huson DH (2008) MetaSim: a sequencing simulator for genomics and metagenomics. PLoS One 3:e3373

Rillig MC, Lehmann A, Lehmann J, Camenzind T, Rauh C (2018) Soil biodiversity effects from field to fork. Trends Plant Sci 23:17–24

Rinke C, Schwientek P, Sczyrba A, Ivanova NN, Anderson IJ et al (2013) Insights into the phylogeny and coding potential of microbial dark matter. Nature 499:431–437

Rinke C et al (2014) Obtaining genomes from uncultivated environmental microorganisms using FACS-based single-cell genomics. Nat Protoc 9:1038–1048

Roesch LF, Fulthorpe RR, Riva A, Casella G, Hadwin AK, Kent AD, Daroub SH, Camargo FA, Farmerie WG, Triplett EW (2007) Pyrosequencing enumerates and contrasts soil microbial diversity. ISME J 1:283–290

Romeh AA, Hendawi MY (2017) Biochemical interactions between *Glycine max* L. silicon dioxide (SiO2) and plant growth-promoting bacteria (PGPR) for improving phytoremediation of soil contaminated with fenamiphos and its degradation products. Pestic Biochem Physiol 142:32–43

Rodriguez-R LM, Konstantinidis KT (2014) Bypassing cultivation to identify bacterial species. Microbe 9(3):111–118

Rosen GL, Reichenberger ER, Rosenfeld AM (2011) NBC: the Naïve Bayes Classification tool webserver for taxonomic classification of metagenomic reads. Bioinformatics 27:127–129

Rothschild LJ, Mancinelli RL (2001) Review article life in extreme environments. Nature 409:1092–1101

Rusch DB, Halpern AL, Sutton G et al (2007) The sorcerer II global ocean sampling expedition: northwest Atlantic through eastern tropical pacific. PLoS Biol 5:e77

Salam JA, Hatha MA, Das N (2017) Microbial-enhanced lindane removal by sugarcane (*Saccharum officinarum*) in doped soil-applications in phytoremediation and bioaugmentation. J Environ Manag 193:394–399

Salzberg SL, Pertea M, Delcher AL, Gardner MJ, Tettelin H (1999) Interpolated Markov models for eukaryotic gene finding. Genomics 59:24–31

Santoyo G, Moreno-Hagelsiebb G, Orozco-Mosquedac MC, Glick BR (2016) Plant growth-promoting bacterial endophytes. Microbiol Res 183:92–99

Sanzani SM, Nicosia MGLD, Faedda R, Cacciola SO, Schena L (2014) Use of quantitative PCR detection methods to study biocontrol agents and phytopathogenic fungi and oomycetes in environmental samples. J Phytopathol 162:1–13

Sarathambal C, Ilamurugu K, Balachandar D, Chinnadurai C, Gharde Y (2015) Characterization and crop production efficiency of diazotrophic isolates from the rhizosphere of semi-arid tropical grasses of India. Appl Soil Ecol 87:1–10

Sasse J, Martinoia E, Northen T (2018) Feed your friends: do plant exudates shape the root microbiome? Trends Plant Sci 23:25–41

Saxena A, Raghuwanshi R, Singh HB (2015) *Trichoderma* species mediated differential tolerance against biotic stress of phytopathogens in *Cicer arietinum* L. J Basic Microbiol 55:195–206

Schadt CW, Zhou J (2005) Advances in microarrays for soil microbial community analyses. In: Nannipieri P, Smalla K (eds) Soil biology: nucleic acids and proteins in soil. Springer-Verlag, New York

Scheinert P, Kruse R, Ullmann U, Söller R, Krupp G (1996) Molecular differentiation of bacteria by PCR amplification of the 16S-23S rRNA spacer. J Microbiol Methods 26:103–117

Schimak MP, Kleiner M, Wetzel S, Liebeke M, Dubilier N, Fuchsa BM (2016) MiL-FISH: multilabeled oligonucleotides for fluorescence in situ hybridisation improve visualisation of bacterial cells. Appl Environ Microbiol 82:62–70

Schneider T, Keiblinger KM, Schmid E, Sterflinger-Gleixner K, Ellersdorfer G, Roschitzki B, Richter A, Eberl L, Zechmeister-Boltenstern S, Riedel K (2012) Who is who in litter decomposition? Metaproteomics reveals major microbial players and their biogeochemical functions. ISME J 6:1749–1762

Schreiber F, Gumrich P, Daniel R, Meinicke P (2010) Treephyler: fast taxonomic profiling of metagenomes. Bioinformatics 26:960–961

Schulze WX, Gleixner G, Kaiser K, Guggenberger G, Mann M, Schulze ED (2004) A proteomic fingerprint of dissolved organic carbon and of soil particles. Oecologia 142:335–343

Schwieger F, Tebbe CC (1998) A new approach to utilize PCR-single-strand-conformation polymorphism for 16S rRNA gene-based microbial community analysis. Appl Environ Microbiol 64:4870–4876

Sebastian R, Kim JY, Kim TH, Lee KT (2013) Metagenomics: a promising approach to assess enzymes biocatalyst for biofuel production. Asian J Biotechnol 5:33–50

Sedlar K, Kupkova K, Provaznik I (2017) Bioinformatics strategies for taxonomy independent binning and visualization of sequences in shotgun metagenomics. Comput Struct Biotechnol J 15:48–55

Segata N, Izard J, Waldron L, Gevers D, Miropolsky L, Garrett WS, Huttenhower C (2011) Metagenomic biomarker discovery and explanation. Genome Biol 12:R60

Segata N, Waldron L, Ballarini A, Narasimhan V, Jousson O, Huttenhower C (2012) Metagenomic microbial community profiling using unique clade-specific marker genes. Nat Methods 9:811–814

Segata N, Boernigen D, Tickle TL, Morgan XC, Garrett WS, Huttenhower C (2013) Computational meta'omics for microbial community studies. Mol Syst Biol 9:666. https://doi.org/10.1038/msb.2013.22

Sekar R, Pernthaler A, Pernthaler J, Warnecke F, Posch T, Amann R (2003) An improved protocol for quantification of freshwater Actinobacteria by fluorescence in situ hybridization. Appl Environ Microbiol 69:2928–2935

Selvakumar N, Ding BC, Wilson SM (1997) Separation of DNA strands facilitates detection of point mutations by PCR-SSCP. BioTechniques 22:604–606

Seshadri R, Kravitz SA, Smarr L, Gilna P, Frazier M (2007) CAMERA: a community resource for metagenomics. PLoS Biol 5:e75

Shah N, Tang H, Doak TG, Ye Y (2011) Comparing bacterial communities inferred from 16S rRNA gene sequencing and shotgun metagenomics. Symp Biocomput Pac:165–176. https://doi.org/10.1142/9789814335058 0018

Shapiro BJ (2017) The population genetics of pangenomes. Nat Microbiol 2(12):1574–1574. https://doi.org/10.1038/s41564-017-0066-6

Sharkey FH, Banat IM, Marchant R (2004) Detection and quantification of gene expression in environmental bacteriology. Appl Environ Microbiol 70:3795–3806

Sharma, CM, Hoffmann S, Darfeuille F, Reignier J, Findeiß S, Sittka A, Stadler PF (2010) The primary transcriptome of the major human pathogen Helicobacter pylori. Nature 464:250–255. https://doi.org/10.1038/nature08756.

Sharon I (2010) Computational methods for metagenomic analysis. Ph.D Thesis, The Technion – Israel Institute of Technology

Sharpton TJ (2014) An introduction to the analysis of shotgun metagenomic data. Front Plant Sci 5:209

Sheng X, Sun L, Huang Z, He L, Zhang W, Chen Z (2012) Promotion of growth and Cu accumulation of bio-energy crop (*Zea mays*) by bacteria: implications for energy plant biomass production and phytoremediation. J Environ Manag 103:58–64

Shokrall SI, Spall JL, Gibson JF, Hajibabaei M (2012) Mol Ecol 21:1794–1805

Sims D, Sudbery I, Ilott NE, Heger A, Ponting CP (2014) Sequencing depth and coverage: key considerations in genomic analyses. Nat Rev Genet 2:121–132

Singh BK, Trivedi P (2017) Microbiome and the future for food and nutrient security. Microb Biotechnol 10:50–53

Singh BK, Campbell CD, Sorenson SJ, Zhou J (2009) Soil genomics. Nat Rev Microbiol 7:756. https://doi.org/10.1038/nrmicro2119-c1

Singh DP, Prabha R, Yandigeri MS, Arora DK (2011) Cyanobacteria-mediated phenylpropanoids and phytohormones in rice (*Oryza sativa*) enhance plant growth and stress tolerance. Antonie Van Leeuwenhoek 100:557–568

Singh A, Dubey PK, Chaurasiya R, Mathur N, Kumar G, Bharati S, Abhilash PC (2018) Indian spinach: an underutilized perennial leafy vegetable for nutritional security in developing world. Energ Ecol Environ. https://doi.org/10.1007/s40974-018-0091-1

Singleton I, Merringto G, Colvan S, Delahunty JS (2003) The potential of soil protein-based methods to indicate metal contamination. Appl Soil Ecol 654:1–8

Sinha S, Mukherjee SK (2008) Cadmium-induced siderophore production by a high Cd-resistant bacterial strain relieved Cd toxicity in plants through root colonization. Curr Microbiol 56:55–60

Smets W, Leff JW, Bradford MA, McCulley RL, Lebeer S, Fierer N (2016) A method for simultaneous measurement of soil bacterial abundances and community composition via 16S rRNA gene sequencing. Soil Biol Biochem 96:145–151

Smith CJ, Osborn AM (2009) Advantages and limitations of quantitative PCR (Q-PCR)-based approaches in microbial ecology. FEMS Microbiol Ecol 67:6–20

Sriprang R, Hayashi M, Ono H, Takagi M, Hirata K, Murooka Y (2003) Enhanced accumulation of Cd2+ by a *Mesorhizobium* sp. transformed with a gene from *Arabidopsis thaliana* coding for phytochelatin synthase. Appl Environ Microbiol 69:1791–1796

Stanley CE, van der Heijden MG (2017) Microbiome-on-a-Chip: new frontiers in plant-microbiota research. Trends Microbiol 25:610–613

Stepanauskas R (2012) Single cell genomics: an individual look at microbes. Curr Opin Microbiol 15:613–620

Sultan M, Amstislavskiy V, Risch T, Schuette M, Dökel S, Ralser M, Balzereit D, Lehrach H, Yaspo ML (2014) Influence of RNA extraction methods and library selection schemes on RNA-seq data. BMC Genomics 15(1):675. http://doi.org/10.1186/1471-2164-15-675

Suominen L, Jussila MM, Mäkeläinen K, Romantschuk M, Lindström K (2000) Evaluation of the Galega–*Rhizobium galegae* system for the bioremediation of oil-contaminated soil. Environ Pollut 107:239–244

Szilagyi-Zecchin VJ, Ikeda AC, Hungria M, Adamoski D, Kava-Cordeiro V, Glienke C, Galli-Terasawa LV (2014) Identification and characterization of endophytic bacteria from corn (*Zea mays* L.) roots with biotechnological potential in agriculture. AMB Express 4:26. https://doi.org/10.1186/s13568-014-0026-y

Tanca A, Palomba A, Pisanu S, Deligios M, Fraumene C, Manghina V, Pagnozzi D, Addis MF, Uzzau S (2014) A straightforward and efficient analytical pipeline for metaproteome characterization. Microbiome 10:49

Tang Y, Underwood A, Gielbert A, Woodward MJ, Petrovska L (2014) Metaproteomics analysis reveals the adaptation process for the chicken gut microbiota. Appl Environ Microbiol 80:478–485

Tao A, Pang F, Huang S, Yu G, Li B, Wang T (2014) Characterisation of endophytic *Bacillus thuringiensis* strains isolated from wheat plants as biocontrol agents against wheat flag smut. Biocontrol Sci Tech 24:901–924

Tara N, Afzal M, Ansari TM, Tahseen R, Iqbal S, Khan QM (2014) Combined use of alkane-degrading and plant growth-promoting bacteria enhanced phytoremediation of diesel contaminated soil. Int J Phytoremediation 16:1268–1277

Tatusov RL, Fedorova ND, Jackson JD, Jacobs AR, Kiryutin B, Koonin EV, Krylov DM, Mazumder R, Mekhedov SL, Nikolskaya AN, Rao BS, Smirnov S, Sverdlov AV, Vasudevan S, Wolf YI, Yin JJ, Natale DA (2003) The COG database: an updated version includes eukaryotes. BMC Bioinform 11:41

Taylor JD, McKew BA, Kuhl A, McGenity TJ, Underwood GJC (2013) Microphytobenthic extracellular polymeric substances (EPS) in intertidal sediments fuelboth generalist and specialist EPS-degrading bacteria. Limnol Oceanogr 58:1463–1480

Teeling H, Glöckner FO (2012) Current opportunities and challenges in microbial metagenome analysis – a bioinformatic perspective. Brief Bioinform 3:728–742

Teeling H, Waldmann J, Lombardot T, Bauer M, Glöckner FO (2004) TETRA: a web-service and a stand-alone program for the analysis and comparison of tetranucleotide usage patterns in DNA sequences. BMC Bioinform 5:163

Teira E, Reinthaler T, Pernthaler A, Pernthaler J, Herndl GJ (2004) Combining catalyzed reporter deposition-fluorescence in situ hybridization and microautoradiography to detect substrate utilization by bacteria and archaea in the deep ocean. Appl Environ Microbiol 70:441–4414

Teixeira C, Almeida CMR, da Silva MN, Bordalo AA, Mucha AP (2014) Development of autochthonous microbial consortia for enhanced phytoremediation of salt-marsh sediments contaminated with cadmium. Sci Total Environ 493:757–765

Telenius H et al (1992) Degenerate oligonucleotide-primed PCR: general amplification of target DNA by a single degenerate primer. Genomics 13:718–725

Tewari S, Arora NK (2013) In: Arora NK (ed) Plant microbe symbiosis: fundamentals and advances, vol 2013. Springer, New Delhi, pp 105–126

Thomas T, Gilbert J, Meyer F (2012) Metagenomics – a guide from sampling to data analysis. Microb Inform Exp 2:3

Torsvik V, Ovreas L (2002) Microbial diversity and function in soil: from genes to ecosystems. Curr Opin Microbiol 5:240–245

Tour JM, Kittrell C, Colvin V (2010) Green carbon as a bridge to renewable energy. Nat Mater 9:871–874

Trapnell C (2015) Defining cell types and states with single-cell genomics. Genome Res 10:1491–1498

Treangen GJ, Koren S, Sommer DD, Liu B, Astrovskaya I, Ondov B, Darling AE, Phillippy AM, Pop M (2013) MetAMOS: a modular and open source metagenomic assembly and analysis pipeline. Genome Biol 14:R2

Tripathi V, Dubey RK, Edrisi SA, Narain K, Singh HB, Singh N, Abhilash PC (2014a) Towards the ecological profiling of a pesticide contaminated soil site for remediation and management. Ecol Eng 71:318–325

Tripathi V, Dubey RK, Singh HB, Singh N, Abhilash PC (2014b) Is *Vigna radiata* (L.) R Wilczek a suitable crop for Lindane contaminated soil? Ecol Eng 73:219–223

Tripathi V, Abhilash PC, Singh HB, Singh N, Patra DD (2015a) Effect of temperature variation on lindane dissipation and microbial activity in soil. Ecol Eng 79:54–59

Tripathi V, Fraceto LF, Abhilash PC (2015b) Sustainable clean-up technologies for soils contaminated with multiple pollutants: plant-microbe-pollutant and climate nexus. Ecol Eng 82:330–335

Tripathi V, Edrisi SA, Abhilash PC (2016a) Towards the coupling of phytoremediation with bioenergy production. Renew Sust Energ Rev 57:1386–1389

Tripathi V, Edrisi SA, O'Donovan A, Gupta VK, Abhilash PC (2016b) Bioremediation for fueling the biobased economy. Trends Biotechnol 34:775–777

Tripathi V, Edrisi SA, Chen B, Gupta VK, Abhilash PC, Vilu R, Gathergood N (2017) Biotechnological advances for restoring degraded land for sustainable development. Trends Biotechnol. https://doi.org/10.1016/j.tibtech.2017.05.001

Trivedi P, Anderson IC, Singh BK (2013) Microbial modulators of soil carbon storage: integrating genomic and metabolic knowledge for global prediction. Trends Microbiol 21:641–651

Troutt AB, McHeyzer-Williams MG, Pulendran B, Nossal GJ (1992) Ligation-anchored PCR: a simple amplification technique with single-sided specificity. Proc Natl Acad Sci U S A 89:9823–9825

Tseng CH, Tang SL (2014) Marine microbial metagenomics: from individual to the environment. Int J Mol Sci 15:8878–8892

Turnbaugh PJ, Ley RE, Hamady M, Fraser-Liggett CM, Knight R, Gordon JI (2007) The human microbiome project. Nature 449(7164):804–810. https://doi.org/10.1038/nature06244

Tveit AT, Urich T, Svenning MM (2014) Metatranscriptomic analysis of arctic peat soil microbiota. Appl Environ Microbiol 80:5761–5772

Tyson GW, Chapman J, Hugenholtz P, Allen EE, Ram RJ, Richardson PM, Solovyev VV, Rubin EM, Rokhsar DS, Banfield JF (2004) Community structure and metabolism through reconstruction of microbial genomes from the environment. Nature 428:37–43

Unlu M, Morgan ME, Minden JS (1997) Difference gel electrophoresis: a single gel method for detecting changes in protein extracts. Electrophoresis 18:2071–2077

van der Heijden MGA, Bardgett RD, van Straalen NM (2008) The unseen majority: soil microbes as drivers of plant diversity and productivity in terrestrial ecosystems. Ecol Lett 11:296–310

Venter JC, Remington K, Heidelberg JF, Halpern AL, Rusch D, Eisen JA, Wu DY, Paulsen I, Nelson KE, Nelson W, Fouts DE, Levy S, Knap AH, Lomas MW, Nealson K, White O, Peterson J, Hoffman J, Parsons R, Baden-Tillson H, Pfannkoch C, Rogers YH, Smith HO (2004) Environmental genome shotgun sequencing of the Sargasso Sea. Science 304:66–74

Verberkmoes NC, Russell AL, Shah M, Godzik A, Rosenquist M, Halfvarson J, Lefsrud MG, Apajalahti J, Tysk C, Hettich RL, Jansson JK (2009) Shotgun metaproteomics of the human distal gut microbiota. ISME J 3:179–189

von Wintzingerode F, Gobel UB, Stackebrandt E (1997) Determination of microbial diversity in environmental samples: pitfalls of PCR-based rRNA analysis. FEMS Microbiol Rev 21:213–229

Walker A, Parkhill J (2008) Single-cell genomics. Nat Rev Microbiol 6:176–177

Wang Z, Chen Y, Li Y (2004) A brief review of computational gene prediction methods. Genomics Proteomics Bioinformatics 2:216–221

Wang W, Vignani R, Scali M, Cresti M (2006) A universal and rapid protocol for protein extraction from recalcitrant plant tissues for proteomic analysis. Electrophoresis 27:2782–2786

Wang L, Feng Z, Wang X, Wang X, Zhang X (2009) DEGseq: an R package for identifying differentially expressed genes from RNA-seq data. Bioinformatics 26:136–138

Wang Y, Leung HC, Yiu SM, Chin FY (2012) MetaCluster 4.0: a novel binning algorithm for NGS reads and huge number of species. J Comput Biol 19:241–249

Wang C, Dong D, Wang H, Müller K, Qin Y, Wang H, Wu W (2016) Metagenomic analysis of microbial consortia enriched from compost: new insights into the role of Actinobacteria in lignocellulose decomposition. Biotechnol Biofuels 9:22

Wang B, Wang Q, Liu W, Liu X, Hou J, Teng Y, Christie P (2017) Biosurfactant-producing microorganism Pseudomonas sp. SB assists the phytoremediation of DDT-contaminated soil by two grass species. Chemosphere 182:137–142

Wang M, Liu P, Xiong W, Zhou Q, Wangxiao J, Zeng Z, Sun Y (2018) Fate of potential indicator antimicrobial resistance genes (ARGs) and bacterial community diversity in simulated manure-soil microcosms. Ecotoxicol Environ Saf 147:817–823

White JR, Navlakha S, Nagarajan N, Ghodsi MR, Kingsford C, Pop M (2010) Alignment and clustering of phylogenetic markers – implications for microbial diversity studies. BMC Bioinform 11:152

White AK et al (2011) High-throughput microfluidic singlecell RT-qPCR. Proc Natl Acad Sci U S A 108:13999–14004

Widjojoatmodjo MN, Fluit AC, Verhoef J (1995) Molecular identification of bacteria by fluorescence-based PCR–single-strand conformation polymorphism analysis of the 16S rRNA gene. J Clin Microbiol 33:2601–2606

Wilke A, Bischof J, Gerlach W, Glass E, Harrison T, Keegan KP, Paczian T, Trimble WL, Bagchi S, Grama A, Chaterji S, Meyer F (2016) The MG-RAST metagenomics database and portal in 2015. Nucleic Acids Res 4:44

Wilkins MR, Sanchez JC, Gooley AA, Appel RD, Humphery-Smith I, Hochstrasser DF, Williams KL (1995) Progress with proteome projects: why all proteins expressed by a genome should be identified and how to do it. Biotechnol Genet Eng Rev 13:19–50

Wilkinson SC, Anderson JM, Scardelis SP, Tisiafouli M, Taylor A, Wolters V (2002) PLFA profiles of microbial communities in decomposing conifer litters subject to moisture stress. Soil Biol Biochem 34:189–200

Williams MA, Taylor EB (2010) Microbial protein in soil: influence of extraction method and C amendment on extraction and recovery. Microb Ecol 59:390–399

Wilmes P, Bond PL (2004) The application of two-dimensional polyacrylamide gel electrophoresis and downstream analyses to a mixed community of prokaryotic microorganisms. Environ Microbiol 6:911–920

Wiśniewski JR, Zougman A, Nagaraj N, Mann M (2009) Universal sample preparation method for proteome analysis. Nat Methods 6:359–362

Witters N, Mendelsohn RO, Van Slyckenc S, Weyens N, Schreurs E, Meers E, Tack F, Carleer R, Vangronsveld J (2012) Phytoremediation, a sustainable remediation technology? Conclusions from a case study. I: energy production and carbon dioxide abatement. Biomass Bioenergy 39:454–469

Wooley JC, Ye Y (2009) Metagenomics: facts and artifacts, and computational challenges. J Comput Sci Technol 25:71–81

Woyke T, Jarett J (2015) Function-driven single-cell genomics. Microb Biotechnol 8:38–39

Wu M, Eisen JA (2008) A simple, fast, and accurate method of phylogenomic inference. Genome Biol. https://doi.org/10.1186/gb-2008-9-10-r151

Wu M, Scott AJ (2012) Phylogenomic analysis of bacterial and archaeal sequences with AMPHORA2. Bioinformatics 28:1033–1034

Wu G, Feng B, Xu J, Zhu XT, Li YC, Zeng NK, Yang ZL (2014) Molecular phylogenetic analyses redefine seven major clades and reveal 22 new generic clades in the fungal family Boletaceae. Fungal Divers 69(1):93–115

Wu C, Wang W, Wang K, Li X, Qiu W, Li W (2016) Phospholipids fatty acids analysis of microbial communities in sewage sludge composting with inorganic bulking agent. Desalin Water Treat 57:27181–27190

Xie X, Liao M, Yang J, Chai J, Fang S, Wang R (2012) Influence of root-exudates concentration on pyrene degradation and soil microbial characteristics in pyrene contaminated soil. Chemosphere 88:1190–1195

Xue K, Wu L, Deng Y, He Z, Nostrand JV, Robertson PG, Schmidt TM, Zhou J (2013) Functional gene differences in soil microbial communities from conventional, low-input, and organic farmlands. Appl Environ Microbiol 79:1284–1292

Xun F, Xie B, Liu S, Guo C (2015) Effect of plant growth-promoting bacteria (PGPR) and arbuscular mycorrhizal fungi (AMF) inoculation on oats in saline-alkali soil contaminated by petroleum to enhance phytoremediation. Environ Sci Pollut Res 22:598–608

Yang M-M, Wen S-S, Mavrodi DV, Mavrodi OV, von Wettstein D, Thomashow LS, Guo J-H, Weller DM (2014) Biological control of wheat root diseases by the CLP-producing strain *Pseudomonas fluorescens* HC1-07. Phytopathology 104:248–256

Yates JR (2004) Mass spectral analysis in proteomics. Annu Rev Biophys Biomol Struct 33:297–316

Yates JR, Speicher S, Griffin PR, Hunkapiller T (1993) Peptide mass maps: a highly informative approach to protein identification. Anal Biochem 214:397–408

Ye Y, Tang H (2009) An ORFome assembly approach to metagenomics sequences analysis. J Bioinforma Comput Biol 7:455–471

Yooseph S, Sutton G, Rusch DB et al (2007) The sorcerer II global ocean sampling expedition: expanding the universe of protein families. PLoS Biol 5:e16

Yu K, Zhang T (2012) Metagenomic and metatranscriptomic analysis of microbial community structure and gene expression of activated sludge. PLoS One 7(5):e38183. https://doi.org/10.1371/journal.pone.0038183

Yu Y, Suh MJ, Sikorski P, Kwon K, Nelson KE, Pieper R (2014) Urine sample preparation in 96-well filter plates for quantitative clinical proteomics. Anal Chem 86:5470–5477

Zaman M, Toth I (2013) Immunostimulation by synthetic lipopeptides based vaccine candidates: structure-activity relationships. Front Immunol 4:1–12

Zarraonaindia I, Smith DP, Gilbert JA (2013) Beyond the genome: community-level analysis of the microbial world. Biol Philos 28:261–282

Zelles L (1999) Fatty acid patterns of phospholipids and lipopolysaccharides in the characterization of microbial communities in soil: a review. Biol Fertil Soils 29:111–129

Zeriouh H, de Vicente A, Pérez-García A, Romero D (2014) Surfactin triggers biofilm formation of *Bacillus subtilis* in melon phylloplane and contributes to the biocontrol activity. Environ Microbiol 16:2196–2211

Zhalnina K, Louie KB, Hao Z, Mansoori N, da Rocha UN, Shi S, Firestone MK (2018) Dynamic root exudate chemistry and microbial substrate preferences drive patterns in rhizosphere microbial community assembly. Nat Microbiol 3:470–480

Zhang MQ (2002) Computational prediction of eukaryotic protein-coding genes. Nat Rev Genet 3:698–709

Zhang L et al (1992) Whole genome amplification from a single cell: implications for genetic analysis. Proc Natl Acad Sci U S A 89:5847–5851

Zhang DY, Brandwein M, Hsuih T, Li HB (2001) Ramification amplification: a novel isothermal DNA amplification method. Mol Diagn 6:141–150

Zhang Q, Zhang J, Yang L, Zhang L, Jiang D, Chen W, Li G (2014) Diversity and biocontrol potential of endophytic fungi in *Brassica napus*. Biol Control 72:98–108

Zhao P, Quan C, Wang Y, Wang J, Fan S (2013) *Bacillus amyloliquefaciens* Q-426 as a potential biocontrol agent against *Fusarium oxysporum* f. sp. spinaciae. J Basic Microbiol. https://doi.org/10.1002/jobm.201200414

Zhou J, Bruns MA, Tiedje JM (1996) DNA recovery from soils of diverse composition. Appl Environ Microbiol 62(2):316–322

Zhou JZ, Thompson DK (2002) Challenges in applying microarrays to environmental studies. Curr Opin Biotechnol 13:204–207

Zhou J, He Z, Yang Y, Deng Y, Tringe SG, Alvarez-Cohen L (2015) High-throughput metagenomic technologies for complex microbial community analysis: open and closed formats. MBio 27:6

Zhou X, Li Z, Zheng T, Yan Y, Li P, Odey EA, Mang HP, Uddin SM (2018) Review of global sanitation development. Environ Int 120:246–261

Zong C, Lu S, Chapman AR, Xie XS (2012) Genomewide detection of single-nucleotide and copy-number variations of a single human cell. Science 338:1622–1626

Zoppini A, Ademollo N, Amalfitano S, Capri S, Casella P, Fazi S, Marxen J, Patrolecco L (2016) Microbial responses to polycyclic aromatic hydrocarbon contamination in temporary river sediments: experimental insights. Sci Total Environ 541:1364–1371

Zornoza R, Acosta JA, Faz A, Baath E (2016) Microbial growth and community structure in acid mine soils after addition of different amendments for soil reclamation. Geoderma 272:64–72

Index

A
Agricultural crops, 13–14
AMPHORA2, 66
Arbuscular mycorrhizal fungi (AMF), 11
Automated version of RISA (Ribosomal
    intergenic spacer analysis (ARISA)
    BLAST, 28
    PCR, 28

B
Bayes classification system, 67
Bioinformatics tools
    FragGeneScan, 65
    Genovo, 61
    green genes, 66
    khmer, 65
    Meta-IDBA toolkit, 65
    MetAMOS, 65
    MetaORFA, 65
    MetaVelvet, 65
    MOCAT, 65
    SOAPdenovo, 65
Bioremediation, 57

C
CAMERA, 70
Carbon sequestration, 17–19, 72
CARMA3, 66
CCREPE, 68
Chemical fertilizers, 2
Classifier for Metagenomic Sequences, 66
Community-level analysis, 44

D
Deferential gradient gel electrophoresis
    (DGGE), 74
    chemical-based denaturation, 27
    cloning and sequencing, 26
    fingerprinting techniques, 26
    herbicide application, 27
    hybridization, 26
    microbial communities, 26
    PCR, 27
DiScRIBinATE, 66
DNA microarrays, 31

E
EBI metagenomics, 70
Ecological suicide, 74
Endophytes, 7
Environmental cleanup technologies, 5
Environmental factors, 25
Environmental microbiology, 26
Extremophiles, 3

F
Fatty acid methyl esters (FAME), 24
Filter-aided sample preparation (FASP), 58
Fluorescence-activated cell sorter (FACS)
    system, 36
Fluorescence in situ hybridization (FISH)
    oligonucleotide, 26
    PAHs, 26
    rRNA-targeted probes, 26
Food security, 3, 17

© The Author(s), under exclusive license to Springer Nature Switzerland AG 2020
R. K. Dubey et al., *Unravelling the Soil Microbiome*, SpringerBriefs in
Environmental Science, https://doi.org/10.1007/978-3-030-15516-2

Functional annotation, 41
Functional profiling, 68

G

Gene calling, 43
General-Error-Model based Simulators, 69
Genomes-OnLine-Database (GOLD), 44
Genomic and proteomic analysis, 4
GeoChip 3.0 microarray technology, 32
Global Ocean Sample (GOS), 43
Glomalin, 15
Greenhouse gas emissions, 4

H

Hexachlorocyclohexane (HCH), 29
Hidden Markov-Model (HMM), 43
High-throughput sequencing technologies,
    46–47
Horseradish peroxidase (HRP) enzyme, 25
HUMAnN, 68

I

IDBA-UD, 69
Induced systemic resistance (ISR), 73
INDUS, 66
Interpolated-Markov-Model (IMM), 43

K

KeggMapper tool, 46, 48
Kyoto-Encyclopedia, 44

L

LefSe, 68
Lowest common ancestor (LCA), 67

M

Marginal lands, 15
MARTA, 66
Metabolic pathways, 68
MetaCluster, 66
MetaGeneAnnotator (MGA), 65
Metagenomic approaches
    applications, 48
    data mining approaches, 47
    EBI, 44
    ecosystems, 39
    gene finding and comparative, 41

interlinking soil, 49
microbial communities, 39
phylogenetic analysis, 42
rRNA gene, 39
Metagenomic data, 39, 45
Metagenomic dataset, 47
Metagenomics, 40
    and metatranscriptomics, 55
Metagenomics data analysis
    AMPHORA, 45
    assembly, 41
    community-level analysis, 44
    comparing pathways, 46
    computational biology, 46
    DNA, 46
    EBI, 44
    gene calling, 43
    high-throughput techniques, 41
    MEGAN, 45
    MG-RAST, 44
    NGS data, 42
    phylogenetic analysis, 42
    sequence binning, 42
    taxonomic binning, 42
Meta-IDBA toolkit, 65
MetaPhlAn, 66
MetaPhyler, 67
METAREP, 70
metaSHARK, 68
MetaSim, 69
Metastats, 68
Metatranscriptomics
    applications, 60
    biochemical approaches, 57
    biological interpretation, 54
    characterization, 56
    deep sequencing, 54
    differential gene expression, 53
    DNA extraction, 56
    environmental conditions, 57
    functional data, 52
    microbial biomass, 59
    mRNA, 51, 53
    probe application, 53
    proteomics, 55
    RNA-seq reads, 53
    soil environment, 59
    2-D gel electrophoresis, 56
Metatranscriptomics data analysis, 61
MG-RAST, 70
Microbial communities, 4
Microbial community structure, 24–26, 28
Microbial diversity, 2

Microbial ecology, 32, 55
Microbial fingerprinting technique, 26
Microbially assisted bioremediation, 20–21
Microbial world, 3
Microorganisms, 73
Molecular-based methodologies, 31
Multiple displacement amplification, 35
Multiple sequence alignment (MSA), 42

N
Next-generation sequencing (NGS), 35, 42
    technologies, 73
N-fixation, 10
Nuclear isolation, 36

O
Operational taxonomic unit (OTU), 28, 75
Orphelia, 65

P
PaPaRa, 67
Phospholipid fatty acid analysis (PLFA), 24
    attribute, 24
    difference, 25
    lipids, 24
    SOM, 25
Phylogenetic analysis, 42
Phymm, 67
Phyto-bioremediation, 72
Plant growth-promoting microorganisms
    (PGPMs), 6
Plant–microbe interactions, 17
Predicted-Relative-Metabolic-Turnover, 68
Prokaryotes, 53
Protein microarray, 58
Proteofingerprint, 57
Pyrosequencing, 52

Q
Quality-control analysis, 66
Quantitative real-time PCR (qPCR), 30

R
RAIphy, 67
RAMMCAP, 68
Rapid annotation by subsystem technology, 46
Restriction enzyme (RE), 27
Rhizospheric microorganisms, 15

Ribosomal RNA (rRNA) genes
    sequencing, 65
Ribosomal-Database-Project, 42, 66
Rice production, 11

S
Sanger sequencing data, 41
Sequence binning, 42
Sequencing single bacterial cells, 35
ShotgunFunctionalizeR, 69
SILVA, 65
Simulators tools, 69
Single-cell genomics (SCG), 36
    advantages, 34
    algorithms and software, 38
    characteristics, 38
    development, 34
    isolation, 36
    metagenomics, 34
    WGA, 37
Single-cell sequencing analysis, 69
Single-strand conformation polymorphism
    (SSCP)
    DGGE/TGGE, 29
    PCR, 29
    16S–23S rRNA, 29
SmashCell, 69
Smash-Community, 70
Soil function, 2
Soil metatranscriptomics, 52
Soil microbial communities, 24, 54, 60, 71
Soil microbial community analysis, 62–64
Soil microorganisms, 3, 6
Soil organic matter (SOM), 25
SOrt-ITEMS, 67
SourceTracker, 69
Spar-CC, 68
SPHINX, 67
Stable isotope probing (SIP), 30
STAMP, 70
Stress tolerance, 7
Suppressive soils, 39
Sustainable development goals (SDGs), 2, 72

T
TACOA, 67
Taxonomic binning, 42, 46
Taxonomic profiling
    AMPHORA2, 66
    CARMA3, 66
    ClaMS, 66

Taxonomic profiling (*cont.*)
    INDUS, 66
    MARTA, 66
    MetaPhlAn, 66
    MetaPhyler, 67
    NBC, 67
    Phymm, 67
    TACOA, 67
    Treephyler, 67
Terminal restriction fragment length
        polymorphism (T-RFLP)
    18S rRNA genes, 27
    microbial populations, 27
    PCR, 27

rRNA sequences, 27
    TRF, 28
Treephyler, 67
2-D gel electrophoresis, 56

V
VAMPS, 70
Volatile organic compounds (VOCs), 12

W
Whole-genome amplification (WGA), 37
    process, 35